Climate Change: A Very Short Introduction

VERY SHORT INTRODUCTIONS are for anyone wanting a stimulating and accessible way into a new subject. They are written by experts, and have been translated into more than 45 different languages.

The series began in 1995, and now covers a wide variety of topics in every discipline. The VSI library currently contains over 650 volumes—a Very Short Introduction to everything from Psychology and Philosophy of Science to American History and Relativity—and continues to grow in every subject area.

Very Short Introductions available now:

ABOLITIONISM Richard S. Newman
THE ABRAHAMIC RELIGIONS
    Charles L. Cohen
ACCOUNTING Christopher Nobes
ADOLESCENCE Peter K. Smith
ADVERTISING Winston Fletcher
AERIAL WARFARE Frank Ledwidge
AESTHETICS Bence Nanay
AFRICAN AMERICAN RELIGION
    Eddie S. Glaude Jr
AFRICAN HISTORY John Parker and
    Richard Rathbone
AFRICAN POLITICS Ian Taylor
AFRICAN RELIGIONS Jacob K. Olupona
AGEING Nancy A. Pachana
AGNOSTICISM Robin Le Poidevin
AGRICULTURE Paul Brassley and
    Richard Soffe
ALEXANDER THE GREAT
    Hugh Bowden
ALGEBRA Peter M. Higgins
AMERICAN BUSINESS HISTORY
    Walter A. Friedman
AMERICAN CULTURAL HISTORY
    Eric Avila
AMERICAN FOREIGN RELATIONS
    Andrew Preston
AMERICAN HISTORY Paul S. Boyer
AMERICAN IMMIGRATION
    David A. Gerber
AMERICAN LEGAL HISTORY
    G. Edward White
AMERICAN MILITARY HISTORY
    Joseph T. Glatthaar

AMERICAN NAVAL HISTORY
    Craig L. Symonds
AMERICAN POLITICAL HISTORY
    Donald Critchlow
AMERICAN POLITICAL PARTIES
    AND ELECTIONS L. Sandy Maisel
AMERICAN POLITICS
    Richard M. Valelly
THE AMERICAN PRESIDENCY
    Charles O. Jones
THE AMERICAN REVOLUTION
    Robert J. Allison
AMERICAN SLAVERY
    Heather Andrea Williams
THE AMERICAN SOUTH
    Charles Reagan Wilson
THE AMERICAN WEST Stephen Aron
AMERICAN WOMEN'S HISTORY
    Susan Ware
AMPHIBIANS T. S. Kemp
ANAESTHESIA Aidan O'Donnell
ANALYTIC PHILOSOPHY
    Michael Beaney
ANARCHISM Colin Ward
ANCIENT ASSYRIA Karen Radner
ANCIENT EGYPT Ian Shaw
ANCIENT EGYPTIAN ART AND
    ARCHITECTURE Christina Riggs
ANCIENT GREECE Paul Cartledge
THE ANCIENT NEAR EAST
    Amanda H. Podany
ANCIENT PHILOSOPHY Julia Annas
ANCIENT WARFARE
    Harry Sidebottom

Available soon:

For more information visit our website

www.oup.com/vsi/

Mark Maslin

CLIMATE CHANGE

A Very Short Introduction

FOURTH EDITION

OXFORD
UNIVERSITY PRESS

OXFORD

UNIVERSITY PRESS

Great Clarendon Street, Oxford, OX2 6DP,
United Kingdom

Oxford University Press is a department of the University of Oxford.
It furthers the University's objective of excellence in research, scholarship,
and education by publishing worldwide. Oxford is a registered trade mark of
Oxford University Press in the UK and in certain other countries

© Eco-Climate Limited 2021

The moral rights of the author have been asserted

First edition published as Global Warming: VSI in 2004
Second edition published as Global Warming: VSI in 2009
Third edition published as Climate Change: VSI in 2014
Fourth edition published as Climate Change: VSI in 2021

Published in the United States of America by Oxford University Press
198 Madison Avenue, New York, NY 10016, United States of America

British Library Cataloguing in Publication Data
Data available

Library of Congress Control Number: 2020951751

ISBN 978-0-19-886786-9

Printed and bound by
CPI Group (UK) Ltd, Croydon, CR0 4YY

Links to third party websites are provided by Oxford in good faith and
for information only. Oxford disclaims any responsibility for the materials
contained in any third party website referenced in this work.

To Chris Pace (1968–2006), Nick Shackleton (1937–2006)
and Anne Maslin (1943–2020)
who never saw problems, only solutions

Contents

Preface to the fourth edition

Climate change is one of the four defining challenges of the 21st century, along with environmental degradation, global inequality, and global insecurity. Climate change will continue to increase the temperature of the Earth and raise global sea level. It will increase the frequency of extreme weather events such as droughts, heat waves, floods, and storms threatening the health and livelihoods of billions of people. The severity of these climate change impacts will depend on what we do now to cut greenhouse gas emissions.

In the past thirty years, the amount of carbon dioxide emitted through human activity has doubled. This represents a collective failure of the world's leaders to focus on the climate crisis. Despite 2020 and 2021 being dominated by the Covid-19 pandemic, the geopolitical landscape around climate change has shifted seismically (Figure 1). In June 2019, the UK parliament amended the 2008 Climate Change Act to requiring that the government reduce the UK's emissions of greenhouse gases to net zero by 2050. In 2021 the UK announced an interim target of a 78% cut in carbon emissions by 2030. The European Commission announced that the EU would reduce its greenhouse gas emissions by at least 55% from 1990 levels by 2030, instead of the 40% cut agreed six years ago. This is a major step towards the EU's overarching pledge of carbon neutrality by 2050. In September 2020, China's President Xi Jinping announced via video-link to the UN General

Assembly in New York that the country would aim to reach peak emissions before 2030, followed by a long-term target to become carbon neutral by 2060. China is the world's largest carbon emitter, accountable for around 28% of global emissions, and up to now has not committed to a long-term emissions goal.

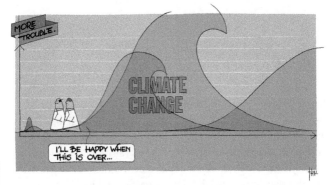

1. **Flattening the curve: comparing Covid-19 and climate change.**

In 2021, the USA, the second largest emitter with around 15% of global emissions, has re-engaged with the climate negotiations. President Trump in 2020 took the USA out of the 2015 Paris Agreement. President Biden has re-engaged The USA in the Paris Agreement and has become a strong advocate of collective international action to deal with climate change. In 2021 the US announced a target cut of 50% of its carbon emissions by 2030 and pledged to reach net carbon zero by 2050. President Biden has also reinstated the environmental regulations removed by President Trump, put in major policies to cut greenhouse gas emissions and greatly increased federal funding for renewable energy and the US green economy. For the first time in over a decade there is now hope that the nations of the world can cut greenhouse gas emissions significantly and start the journey to a cleaner, greener, safer, healthier, and more sustainable world.

Acknowledgements

The author would like to thank the following people: Johanna, Alexandra, and Abbie for surviving lockdown together and allowing me to write the fourth edition; Miles Irving for his excellent illustrations; my editors at OUP Jenny Nugee and Latha Menon; all the wonderful staff at UCL, Sopra Steria Group, Sheep Included, The Conversation, and Rezatec Ltd; Richard Betts, Mark Brandon, Andrew Shepherd, Eric Wolff, and other reviewers, for their insightful and extremely helpful reviews of the different editions of this book; and all my brilliant, dedicated colleagues in climatology, palaeoclimatology, geology, geography, social science, economics, medicine, engineering, humanities, and the arts who continue to strive to understand, predict, and mitigate our influence on the climate of our planet.

List of illustrations

Abbreviations

AABW	Antarctic Bottom Water
AO	Arctic oscillation
AOGCM	atmosphere–ocean general circulation model
AOSIS	Alliance of Small Island States
AR	IPCC Assessment Report
CCS	carbon capture and storage
CDM	Clean Development Mechanism
CFCs	chlorofluorocarbons
CH_4	methane
CMA3	third meeting of the parties to the Paris Agreement
CMIP	Coupled Model Intercomparison Projects
CO_2	carbon dioxide
COP	Conference of the Parties
ECS	equilibrium climate sensitivity
ENSO	El Niño–Southern Oscillation
ETS	Emissions Trading Scheme
G8	Group of Eight
GCM	general circulation model
GCR	galactic cosmic ray
GDP	gross domestic product
GHCN	Global Historical Climate Network
GHG	greenhouse gas
GMT	global mean temperature

Gt	gigatonnes
GtC	gigatonnes of carbon
HCO_3-	calcium bicarbonate
IEA	International Energy Agency
IMF	International Monetary Fund
IPCC	Intergovernmental Panel on Climate Change
MAT	marine air temperature
MW	megawatts
NADW	North Atlantic Deep Water
NAO	North Atlantic Oscillation
NASA	National Aeronautics and Space Administration
NGO	non-governmental organization
N_2O	nitrous oxide
NOAA	National Oceanic and Atmospheric Administration
NRC	National Research Council (USA)
NSA	National Security Agency (USA)
OECD	Organization for Economic Cooperation and Development
OPEC	Organization of the Petroleum Exporting Countries
PETM	Palaeocene–Eocene Thermal Maximum
ppbv	parts per billion by volume
ppmv	parts per million by volume
pptv	parts per trillion by volume
RCP	representative concentration pathway
REDD+	Reduced Emissions from Deforestation and Forest Degradation, including safeguards for local people
SRES	Special Report on Emission Scenarios by the IPCC (2000)
SSP	shared socioeconomic pathway
SST	sea-surface temperature
UNCED	United Nations Conference on Environment and Development

UNFCCC	United Nations Framework Convention on Climate Change
VBD	vector-borne disease
WHO	World Health Organization
WTO	World Trade Organization

Chapter 1
What is climate change?

Future climate change is one of the defining challenges of the 21st century, along with global inequality, environmental degradation, and global insecurity. The problem is that 'climate change' is no longer just a scientific concern, but encompasses economics, sociology, geopolitics, national and local politics, law, and health, just to name a few. This chapter will examine the role of greenhouse gases (GHGs) in moderating past global climate, why they have been rising since the Industrial Revolution, and why they are now considered dangerous pollutants. It will examine which countries have produced the most anthropogenic GHGs and how this is changing with rapid economic development. It will introduce the Intergovernmental Panel on Climate Change (IPCC) and how it regularly collates and assesses the most recent evidence for climate change.

The Earth's natural greenhouse

The temperature of the Earth is determined by the balance between energy received from the Sun and its loss back into space. The Sun's energy consists of short-wave radiation (mainly visible 'light' and ultraviolet (UV) radiation) and nearly all of it passes through the atmosphere without interference (see Figure 2). The only exception is damaging high-energy UV, which is absorbed by atmospheric ozone. About one-third of the solar energy is reflected straight back into space. The remaining energy is absorbed by the surface of the Earth. This energy warms the land and the oceans, and this heat is radiated back as long-wave infrared or 'heat' radiation.

Atmospheric gases such as water vapour, carbon dioxide (CO_2), methane (CH_4), and nitrous oxide (N_2O) are known as GHGs as they absorb some of this long-wave radiation, warming the atmosphere. This effect has been measured in the atmosphere and can be reproduced time and time again in the laboratory. Without this natural greenhouse effect the Earth would be at least 35°Celsius (C) colder, making the average temperature in the tropics about −10°C. Since the Industrial Revolution, we have been burning fossil fuels (oil, coal, and natural gas) deposited hundreds of millions of years ago, releasing the carbon back into the atmosphere as CO_2 and CH_4, increasing the 'greenhouse effect', and elevating the temperature of the Earth. In effect we are burning fossilized sunlight.

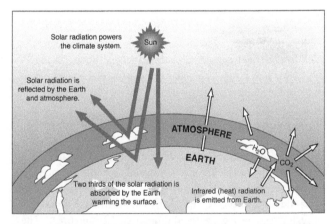

2. **The greenhouse effect. Greenhouse gases trap some of the Earth's heat before releasing it to warm the atmosphere.**

Past climate

Climate change in the geological past has been reconstructed using a number of key archives, including marine and lake sediments, ice cores, cave deposits, and tree rings. These various records reveal that over the past 50 million years the Earth's climate has been cooling down, moving from the so-called 'greenhouse world' of the Eocene, with warm and gentle conditions, through to the cooler and more dynamic 'ice house world' of today. It may seem odd that in geological terms our planet is extremely cold, while this whole book is concerned with our rapid warming of the planet. This is because the very fact that there are huge ice sheets on both Antarctica and Greenland, and nearly permanent sea ice in the Arctic Ocean, makes the global climate very sensitive to changes in GHGs.

The long-term global cooling of the Earth kicked off with the glaciation of Antarctica about 35 million years ago and then accelerated with the great Northern Hemisphere ice ages, which began 2.5 million years ago. Since the beginning of the great ice ages, the global climate has cycled between conditions that were similar or even slightly warmer than today, to full glacial phases, during which ice sheets over 3 kilometres (km) thick formed over much of North America and Europe. Between 2.5 and 1 million years ago, these glacial–interglacial cycles occurred every 41,000 years, and since 1 million years ago they have occurred every 100,000 years.

These great ice-age cycles are driven primarily by changes in the Earth's orbit with respect to the Sun. In fact, the world has spent over 80% of the past 2.5 million years in conditions colder than the present. Our present interglacial, the Holocene Period, began about 10,000 years ago, and is an example of the brief warm conditions that occur between each ice age. The Holocene began with the rapid and dramatic end of the last ice age: in less than

4,000 years global temperatures increased by 6°C, global sea level rose by 120 metres (m), atmospheric CO_2 increased by one-third, and atmospheric CH_4 doubled.

Still, this is much slower than the changes we are seeing today. James Lovelock, in his book *The Ages of Gaia*, suggests that interglacials like the Holocene are the fevered state of our planet, which clearly over the past 2.5 million years prefers a colder average global temperature. Lovelock sees global warming as humanity adding to the already fevered state of the planet. These large-scale past changes in global climate are discussed in more detail in *Climate: A Very Short Introduction*.

Past variations in carbon dioxide

One of the pieces of scientific evidence that shows that atmospheric CO_2 is an important control of global climate comes from the study of past climate. Evidence for past variations in GHGs and temperature comes from ice cores drilled in both Antarctica and Greenland. As snow falls, it is light and fluffy and contains a lot of air. As more snow falls the older snow is slowly compacted to form ice, with some of the air bubbles becoming trapped. By extracting the air from these bubbles trapped in the ancient ice, scientists can measure the percentage of GHGs that were present in the past atmosphere. Scientists have drilled down over 2 miles into both the Greenland and the Antarctic ice sheets, which has enabled them to reconstruct the amount of GHGs present in the atmosphere over the past million years. By examining the oxygen and hydrogen isotopes in the frozen water that make up the ice core, it is possible to estimate the air temperature above the ice sheet when the water first froze.

The results are striking: the proportion of GHGs such as atmospheric CO_2 and CH_4 co-vary with temperature over the past 800,000 years (see Figure 3). The cyclic changes in climate from glacial to interglacial periods can be seen both in temperatures and the GHG content of the atmosphere. This strongly supports the idea that GHGs in the atmosphere and global temperature are closely linked; when CO_2 and CH_4 increase, global temperatures increase, and vice versa when they decrease.

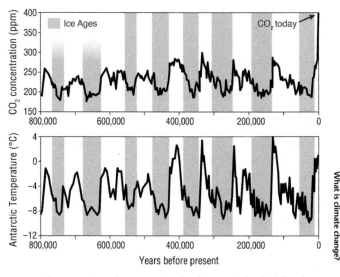

3. **Greenhouse gases and temperature for the past eight glacial cycles recorded in ice cores.**

Early farmers

The high-resolution ice-core evidence from Greenland and the continental margins of Antarctica shows that GHGs in the atmosphere rose a small amount before the Industrial Revolution in the 1700s. Bill Ruddiman, Professor of Palaeoclimatology at the University of Virginia, suggested that early agriculturalists caused this reversal in natural decline in GHGs. Deforestation and land clearing caused the rise in atmospheric CO_2 starting about 7,000 years ago, while the expansion of wet rice agriculture and cattle caused atmospheric CH_4 to start rising about 5,000 years ago.

5

It seems that early human interactions with our environment increased atmospheric GHGs just enough that even prior to the Industrial Revolution we had already delayed the onset of the next ice age, which would otherwise have started gently to occur any time in the next 1,000 years.

The Industrial Revolution

There is clear evidence that levels of atmospheric CO_2 have been rising ever since the beginning of the Industrial Revolution. The first measurements of CO_2 concentrations in the atmosphere started in 1958, on the summit of Mauna Loa Mountain in Hawaii at an altitude of about 4,000 m. The measurements were made at this remote location to avoid contamination from local pollution sources. The record clearly shows that atmospheric concentrations of CO_2 have increased every single year since 1958. The mean concentration of approximately 316 parts per million by volume (ppmv) in 1958 has risen to over 420 ppmv today (see Figure 4). The annual variations in the Mauna Loa observatory are mostly due to CO_2 uptake by growing plants. The uptake is highest in Northern Hemisphere springtime due to the great expanse of land and hence every spring there is a drop in atmospheric CO_2, which unfortunately does nothing to change the overall trend towards ever higher values.

The Mauna Loa observatory CO_2 data can be combined with the detailed ice-core evidence to produce a complete record of atmospheric CO_2 since the beginning of the Industrial Revolution. This shows that atmospheric CO_2 has increased from a pre-industrial concentration of about 280 ppmv to over 420 ppmv at present, representing an increase of over 45%. To put this increase into context, ice-core evidence shows that over the last 800,000 years the natural change in atmospheric CO_2 has been between about 200 and 280 ppmv. The variation between warm and cold periods is about 80 ppmv—less than the CO_2 pollution that we have put into the atmosphere over the past 100 years.

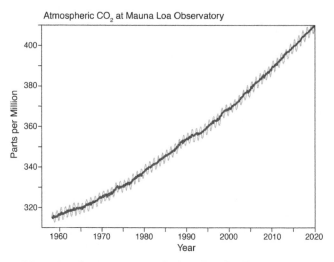

4. Mauna Loa observatory atmospheric carbon dioxide measurements.

The level of human pollution caused in one century has been greater than natural variations, which took thousands of years.

Who produces the pollution?

The United Nations Framework Convention on Climate Change (UNFCCC) was created to produce the first international agreement on reducing global GHG emissions. This is not a simple task as CO_2 emissions are not evenly produced by countries. According to the IPCC (see Box 1) the primary source of CO_2 is the burning of fossil fuels: over 85% of the global CO_2 emissions comes from energy production, industrial processes, and transport. These emissions are not evenly distributed around the world because of the unequal distribution of industry and wealth: North America, Europe, and Asia emit over 90% of the global, industrially produced CO_2 (see Figure 5). Moreover, historically, the developed nations have emitted much more than less developed countries.

Box 1 What is the IPCC?

The IPCC was established in 1988 jointly by the United Nations Environmental Panel and the World Meteorological Organization to address concerns about global warming. The purpose of the IPCC is the continued assessment of the state of knowledge on the various aspects of climate change, including scientific, environmental, and socioeconomic impacts and response strategies. The IPCC does not undertake independent scientific research, rather it brings together all key research published in the world and produces a consensus. There have been six main IPCC reports—in 1990, 1996, 2001, 2007, 2013/14, and 2021/2—and many individual, specialized reports on such subjects as carbon-emission scenarios, alternative energy sources, oceans, land use, and extreme weather events.

The IPCC is recognized as the most authoritative scientific and technical voice on climate change, and its assessments have had a profound influence on the negotiators of the UNFCCC. The IPCC is organized into three working groups plus a task force to calculate the amount of GHGs produced by each country. Each of these four bodies has two co-chairs (one from a developed and one from a developing country) and a technical support unit. Working Group I assesses the scientific aspects of the climate system and climate change; Working Group II addresses the vulnerability of human and natural systems to climate change, and options for adapting to climate change; and Working Group III assesses options for limiting GHGs emissions and otherwise mitigating climate change.

The IPCC provides governments with scientific, technical, and socioeconomic information relevant to evaluating the risks and to developing a response to global climate change. The latest reports from these three working groups were published in 2021—with approximately 500 experts, from some 120 countries, directly involved in drafting, revising, and finalizing the

IPCC reports, as well as thousands of additional experts participating in the review process. The IPCC authors are always nominated by governments and international organizations, including non-governmental organizations (NGOs). These reports are essential reading for anyone interested in climate change and are listed in the Further Reading section at the end of this book. In 2008, the IPCC was jointly awarded, with Al Gore, the Nobel Peace Prize, to acknowledge all the work the IPCC had done over the previous 20 years.

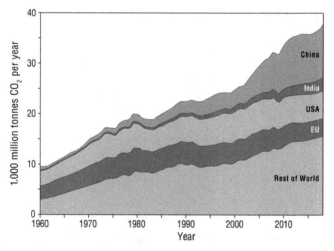

5. **Historic carbon dioxide emissions by region.**

The second major source, accounting for 10–15% of global CO_2 emissions, is land-use changes. These emissions come primarily from the cutting down of forests for the purposes of agriculture, urbanization, or building roads. When rainforest is cut down the land often turns into less productive grassland with considerably reduced capacity for storing CO_2. Here the pattern of CO_2 emissions is different, with South America, Asia, and Africa being responsible for over 90% of present-day land-use change emissions. This raises important ethical questions because it is difficult to tell these countries to stop deforestation when this has already occurred in much of North America and Europe before the beginning of the 20th century. In terms of the amount of CO_2 released, industrial processes still significantly outweigh land-use changes.

We have put nearly half a trillion tonnes of carbon into the atmosphere since the Industrial Revolution, but this still amounts to only half of our total emissions. The other half has been absorbed by the Earth—with 25% going into the oceans and 25% going into the land biosphere. Scientists are concerned that this uptake of our pollution is unlikely to continue at the same level in the future. This is because as global temperatures rise the oceans will warm and will be able to hold less dissolved CO_2. As we continue to deforest and convert land for farming and urbanization, there will be less vegetation to absorb CO_2, again reducing the uptake of our carbon pollution (Figure 6).

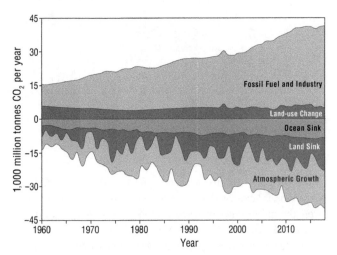

6. **Historic global sinks and sources of carbon dioxide.**

There is clear evidence that GHG concentrations in the atmosphere have been rising since the Industrial Revolution of the 18th century. Atmospheric concentrations of both CO_2 and CH_4 are higher now than at any time within at least the past 3 million years. Within 100 years, we have put more than one and a half times the amount of carbon into the atmosphere than was emitted over the 4,000 years' transition from the last ice age to our current interglacial period.

The current scientific consensus is that these recent changes in GHG concentrations in the atmosphere have already caused global temperatures to increase. Since 1880, the global average surface temperature has increased by 1.1°C. This warming has been accompanied by a significant warming of the ocean, a rise in sea level of over 24 centimetres (cm), a 50% decline in Arctic sea ice, and an increase in the number of extreme weather events. As we emit more and more carbon into the atmosphere, the effects in terms of climate change will increasingly threaten and challenge human society.

The science, politics, and potential solutions to climate change are examined in the rest of this book. In Chapter 2 the emergence of climate change as a global issue is discussed. Chapters 3 and 4 consider the current scientific evidence for climate change, and how scientists are modelling the future to assess the ways in which global carbon emissions will alter our climate. Chapters 5 and 6 examine the impacts of these future climate changes and the possibility that there may be hidden surprises within the climate system that may exacerbate climate change. Chapters 7 and 8 investigate the political aspects of climate change, and the potential political, economic, and technological solutions available to us. Finally, Chapter 9 provides multiple views of the future, dependent on our future carbon emissions, and discusses how we could resolve the climate change crisis.

Chapter 2
History of climate change

Scientists predict that if we continue on our current carbon emissions pathway we could warm the planet by between 1.5°C and 4.7°C in the next 80 years, which economists suggest could cost us as much as 20% of world gross domestic product (GDP). In the face of such a threat, it is crucial to understand the history of climate change and the evidence that supports it. The essential science of climate change was all there in the late 1950s, but it was not taken seriously until the late 1980s. Since then climate change has emerged as one of the biggest scientific and political issues facing humanity.

An old science

The history of climate change science is a long one and can be said to have started in 1856 when Eunice Newton Foote (American scientist, inventor, and women's rights campaigner) published her paper demonstrating the greenhouse effect of CO_2. She used glass cylinders and mercury thermometers, and showed that when they were filled with different gases and placed in direct sunlight, the one containing CO_2 trapped the most heat. Looking to the history of the Earth, Foote theorized: 'An atmosphere of that gas would give to our Earth a higher temperature.'

Just three years later, John Tyndall, professor of natural philosophy at the Royal Institution in London, demonstrated and measured the greenhouse effects of different gases. Using apparatus that utilized thermopile technology, he was the first to correctly measure the relative infrared (heat) absorption of gases such as nitrogen, oxygen, water vapour, CO_2, CH_4, and ozone. He concluded that water vapour is the strongest absorber of radiant heat in the atmosphere and is the principal gas controlling the air temperature of the Earth.

Building on the prior work of scientists such as Tyndall, Joseph Fourier, and Claude Pouillet, in 1896 the Swedish physical chemist Svante Arrhenius calculated how much the Earth's temperature would change given variations in GHGs. He estimated that a halving of atmospheric CO_2 would drop the Earth's temperature by 4°C, and that this may have been a key cause of the ice ages, while a doubling of CO_2 would increase the global temperature by 4°C. He concluded that anthropogenic CO_2 emissions resulting from the burning of fossil fuels would be great enough to cause global warming.

But it wasn't until 1938 that the engineer and inventor Guy Stewart Callendar compiled 147 temperature records from around the world covering the previous 50 years and showed that the world was indeed warming (Figure 7). By using the few atmospheric CO_2 measurements available he was able to suggest that a doubling of atmospheric CO_2 would cause a 2°C warming, half the figure suggested by Arrhenius. Callendar's results were initially met with scepticism, but his papers, published through the 1940s and 1950s, prompted other scientists to investigate changes in atmospheric CO_2 and what was controlling them.

The Second World War saw a massive improvement in technology, including the development of infrared spectroscopy, and soon after the war scientists were able to show that CO_2 in the upper

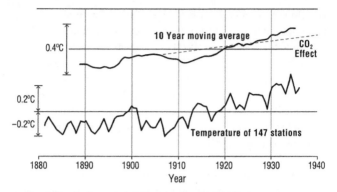

7. Guy Callendar's 1938 global temperature compilation.

atmosphere at low pressure did absorb heat—thus demonstrating the greenhouse effect. The worry about possible global warming was ignored at first as scientists argued that the oceans would simply absorb any extra anthropogenic CO_2 emitted. Roger Revelle, director of the Scripps Institute of Oceanography in California, was concerned by this dismissal. Through his studies of surface ocean chemistry, he found that oceans return much of the CO_2 that they absorb back into the atmosphere. This was a great revelation, and showed that because of the peculiarities of ocean chemistry, the oceans would not be the complete sink for anthropogenic CO_2 as first thought. We now know that the oceans are taking up about one-quarter of the annual total anthropogenic production (Figure 6).

Charles Keeling, who was hired by Revelle, made the next important step forward in climate change science. In the late 1950s and early 1960s, Keeling used the most modern technology available to measure the concentration of atmospheric CO_2 in the Antarctica and Mauna Loa. The resulting Keeling CO_2 curves have continued to climb ominously each year since his first measurement in 1958 and have become one of the major iconic images illustrating global warming (Figure 4).

14

Why the delay in recognizing climate change?

In 1959, the physicist Gilbert Plass published an article in *Scientific American* declaring that the world's temperature would rise by 3°C by the end of the century. The magazine editors published an accompanying photograph of coal smoke belching from factories and the caption read, 'Man upsets the balance of natural processes by adding billions of tons of carbon dioxide to the atmosphere each year.' This resembles thousands of magazine articles, television news items, and documentaries that we have all seen since the late 1980s. So why was there a delay between the science of global warming being accepted in the late 1950s and the realization by those outside the scientific community of the true threat of global warming at the beginning of the 21st century?

The key reasons for the delay in recognizing climate change were the lack of increase in global temperatures and the lack of global environmental awareness. The global mean temperature (GMT) data set is calculated by compiling all the available land and sea temperatures. From 1940 until the mid-1970s, the global temperature curve seems to have had a slight downward trend (Figure 8). This provoked many scientists to discuss whether the Earth was entering the next great ice age. Increasing knowledge of past climates in the 1970s and 1980s showed that this was highly unlikely, as ice ages take thousands of years to form.

Even so, it was not until the late 1980s, when the global annual mean temperature curve started to rise, that the global cooling scenario was finally discarded. By the late 1980s, the global annual mean temperature curve was rising so steeply that all the earlier evidence from the late 1950s and 1960s was dusted off and the global warming theory came to prominence. In fact, it was in 1988 that Professor Jim Hansen, the director of the NASA Goddard Institute for Space Studies, was asked to testify on the matter before the US Senate Committee on Energy and Natural

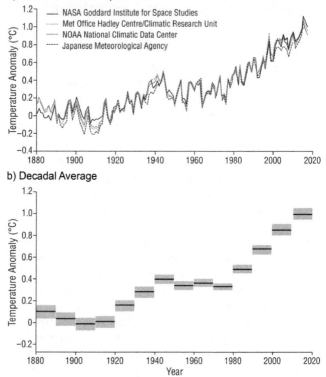

a) Global Surface Temperatures

b) Decadal Average

8. Variation of the Earth's surface temperature over the last 150 years.

Resources. He stated that, 'Global warming has reached a level such that we can ascribe with a high degree of confidence a cause and effect relationship between the greenhouse effect and observed warming...It is already happening now.' This testimony was widely reported by the media, and global warming became a mainstream issue.

It seems, then, that the eventual recognition of climate change was driven by the upturn in the global annual mean temperature.

The latest IPCC 2021 science report has reviewed and synthesized a wide range of data sets. It shows that the trend in global temperature first recognized in the late 1980s is correct, and that this warming trend has continued unstopped until the present day (see Figure 8).

The upturn in the recorded global annual mean temperature was not the sole reason for the new prominence of the global warming issue. In the late 1970s and 1980s, there were significant advances in global climate modelling. These new atmosphere–ocean general circulation models (AOGCMs) produced estimates of significant warming associated with a doubling of CO_2 in the atmosphere—closer in fact to Arrhenius's original calculations. By the 1980s, scientific concern had emerged about CH_4 and other non-CO_2 GHGs, as well as the role of the oceans as a carrier of heat. GCMs continued to improve, and the numbers of scientific teams working on such models increased over the 1980s and the 1990s. In 1992, a first overall comparison of results from fourteen GCMs was undertaken; the results were all in rough overall agreement, confirming that rising GHGs would cause significant global warming.

The rise of the global environmental social movement

The 1980s saw a massive grass-roots expansion in the environmental movement, particularly in the USA, Canada, and the UK, partly as a backlash against the right-wing governments of the 1980s and the expansion of the consumer economy; and partly because of the increasing number of environment-related stories in the media. This heralded a new era of global environmental awareness and the emergence of transnational NGOs. The roots of this growing environmental awareness can be traced back to a number of key markers: these include the publication of Rachel Carson's *Silent Spring* in 1962; the image of Earth seen from the Moon in 1969; the Club of Rome's 1972 report on *Limits to Growth*; the Three Mile Island nuclear reactor accident in 1979; the nuclear accident at Chernobyl in 1986; and the Exxon Valdez oil spillage in 1989 (although the latter three created environmental problems that were all regional in effect, limited geographically to the specific areas in which they occurred).

It was the discovery in 1985 by the British Antarctic Survey of the depletion of ozone over Antarctica that demonstrated the global connectivity of our environment. The ozone 'hole' also had a tangible international cause—the use of chlorofluorocarbons (CFCs)—which led to a whole new area of politics: the international management of the environment. There followed a set of key agreements: the 1985 Vienna Convention for the Protection of the Ozone Layer; the 1987 Montreal Protocol on Substances that Deplete the Ozone Layer; and the 1990 London and 1992 Copenhagen Adjustments and Amendments to the Protocol. These have been held up as examples of successful environmental diplomacy.

These new global environmental concerns and the ability to deal with them internationally were encouraged and articulated by leading politicians of the time. Margaret Thatcher, the prime minster of the UK in 1989, gave an address to the UN outlining the science of climate change, the threat it posed to all nations, and the action needed to avert the crisis. She summed up by saying, 'We should work through this great organisation and its agencies to secure world-wide agreements on ways to cope with the effects of climate change, the thinning of the ozone layer, and the loss of precious species.' George Bush Senior, president of the USA, gave similar speeches, including one in 1992 when he outlined his clear skies and global climate change initiatives at the National Oceanic and Atmospheric Administration.

The IPCC was set up in 1988 and produced its very first science report in 1990. Two years later, with the support of leaders from all around the world, the UN held the Rio Earth Summit, officially called the United Nations Conference on Environment and Development (UNCED), to help member states cooperate on sustainability and protecting the world's environment. The summit was a huge success and led to the Rio Declaration on Environment and Development, the local sustainability initiative called Agenda 21, and Forest Principles. It also set up the United Nations Convention to Combat Desertification, the Convention on Biological Diversity, and the UNFCCC that underlies the negotiations to limit global GHG emissions. The Rio Earth Summit also laid the foundations for the Millennium Development Goals and the subsequent Sustainable Development Goals.

The economists wade in

Economists have been involved with studying climate change from the very beginning of the IPCC process. Two particular publications from economists have had very different effects on the climate change debate. First, there was the publication of the controversial book *The Skeptical Environmentalist* by Bjørn Lomborg in English in 2001. In this and subsequent publications, he argues that the cost of cutting global GHG emissions is extremely high and that those who suffer most are the poorest, so instead we should alleviate poverty by rapidly developing poor countries.

There are two major issues with this approach. First, the costs of switching to a low-carbon economy are relatively low and may even benefit economic growth. Second, it is deeply unrealistic to expect that rich nations will transfer funds to poorer countries on the scale needed to alleviate poverty just to avoid having to cut GHG emissions.

The second major landmark was the publication of the UK government-commissioned, 2006 Stern Report on *The Economics of Climate Change* (published in 2007). The report was led by Sir

Nicholas Stern, then-adviser to the UK government on the economics of climate change and development, reporting to Prime Minister Tony Blair. The report states that if we do nothing, the impacts of climate change could cost between 5% and 20% of world GDP every year. That means the whole world loses one-fifth of what it earns to address the impacts (discussed in Chapter 5).

This of course puts climate change impacts on a completely different economic scale than was envisaged by Lomborg. But the Stern Report does present some good news, arguing that if we do everything we can to reduce global GHG emissions and ensure we adapt to the coming effects of climate change, this will cost us only 1% of world GDP every year.

The Stern Report was criticized by other economists—for example, does it use the right inherent discount rate? The inherent discount rate is the rate economists use to take into account that consumption inherently has a lower value in the future than in the present. In other words, future consumption should be discounted simply because it takes place in the future and people generally prefer the present to the future. The Nobel Laureate William Nordhaus used inherent discount rates of up to 3%, arguing that people today value an environmental benefit that will occur 25 years in the future half as much as they value having that same benefit today.

Nordhaus, however, has recently come under intense criticism as he claims a 4°C increase in global temperature over pre-industrial levels would only reduce GDP per capita by between 2% and 4%. But the fundamental flaw in Nordhaus' model is that he uses a linear not a quadratic damage function—so even catastrophic levels of climate change do not do much economic damage in this economic model. The Stern Report has also been criticized for being overly optimistic about the costs of adapting to a low-carbon world. In June 2008, Stern did revise his estimated costs up to 2%

of world GDP. Nevertheless, the Stern Report sent seismic waves around the world. It was as if people said to themselves, 'If the economists are worried about the cost of climate change, it must be real.'

This was not the end of the involvement of economists in climate change and there have been a number of highly influential books and papers which question our whole understanding of economics and its relationship with the environment. These include the economist Tim Jackson's book *Prosperity without Growth*, first published in 2009, that challenges the orthodox view that economic growth is required or even desirable. In 2017, economist Kate Raworth published *Doughnut Economics*, in which she shows seven ways 'classical economics' has got it wrong, arguing that environmental limits and basic human rights must be at the centre of economics. For the first time for two generations classical economics is under sustained attack from a new generation of dynamic innovative 21st-century economists who see environmental and human wellbeing as integral parts of the world economy. At the centre of this is how we deal with climate change while improving peoples' lives.

Climate change and the media

The other reason for the emergence of climate change as a major global issue was the intense media interest. This is because climate change is perfect for the media: a dramatic story about the end of the world as we know it with key protagonists arguing that it is not even real. The majority of newspaper articles in the UK, the USA, and Australia in the 1990s cast doubt on the claims of climate change. There was a recurrent attempt to promote mistrust in science, through strategies of generalization, of exaggerating disagreement within the scientific community, and, most importantly, discrediting scientists and scientific institutions.

There are two possible explanations for this extraordinarily media-facilitated public scientific debate. First, climate change deniers and industrial lobby groups who do not want to see political action to address climate change are using this debate about methods and scientific uncertainty as a convenient hook on which to hang their case for delay. In fact it has been found that in 2019, five of the largest publicly listed oil companies spent over $200 million lobbying to control, delay, or block binding climate policy.

Second, the media's ethical commitment to balanced reporting, inappropriately applied, draws unwarranted attention to critical views when they are marginal and outside the realm of what is normally considered 'good' science. In the UK, the BBC came under increasing criticism for continually presenting this false balance, usually pitching a climate scientist against a seasoned politician or a paid lobbyist.

Beyond conventional media, the so-called debate about climate change has moved on to social media, with climate change deniers attacking the evidence and views of scientists whenever they can. This rise of fake news has impacted many areas of science, including vaccinations and efforts to tackle Covid-19 as well as climate change. Together, false balance in media debate, fake news, and social media campaigns contribute to a public impression that the science of climate change is 'contested', despite what many would argue is an overwhelming scientific case that climate change is occurring and that human activity is its main driver.

But things are changing and in the past few years in many countries opinion polls have shown that the majority of the public have realized that climate change is real and a major threat. This is mainly through people's own experience or watching the effects of extreme weather around the world. There are now regular news stories about climate change, and they continued even during the recent Covid-19 pandemic. Major documentaries such as Al Gore's *An Inconvenient Truth*, David Attenborough's *A Life on Our Planet*, and BBC1's *Climate Change: The Facts* have also drawn widespread attention to the issue.

The new global environmental social movements

In 2008 and 2009, there was a second global rise in social awareness of climate change. This time it was focused on the hope of a major climate deal at the Copenhagen climate conferences. The Copenhagen conference ended in abject failure due to a lack of international leadership, sabotage by the US, and global worries about dealing with the 2008 global financial crash. It took until the Paris 2015 climate conference to get the negotiation back on track. For almost ten years the environmental movement was held back due to the focus on the global economy. This all changed in 2018, when the third wave of the global environmental social movement began.

In May 2018, the protest group Extinction Rebellion was set up in the UK and launched in October 2018 with over a hundred academics calling for action on climate change. The aim of Extinction Rebellion is to use non-violent civil disobedience to compel governments around the world to avoid tipping points in the climate system and biodiversity loss to prevent both social and ecological collapse. In November 2018 and April 2019, they brought central London to a standstill, and Extinction Rebellion's membership has now spread to at least sixty other cities around the world.

In August 2018, the 15-year-old Greta Thunberg started to spend her school days outside the Swedish parliament holding a sign saying *Skolstrejk för klimatet* (School strike for climate), calling for stronger action on climate change. The message spread. Soon other students all around the world started similar school strikes once a month on a Friday, and they called the movement 'Fridays for Future'. It has been estimated that by the end of 2019 there had been over 4,500 strikes across over 150 countries, involving 4 million schoolchildren.

In 2018 and 2019, three extremely influential IPCC reports were published. The first, in 2018, was the Special Report on Global Warming of 1.5°C, which documented what the world needed to do if global temperature rise was to be kept at only 1.5°C. It also showed the positive and negative interactions of climate change mitigation and the sustainability development goals. The second was the Special Report on the Land, on how climate change would impact desertification, land management, food security, and terrestrial ecosystems. The third was the IPCC Special Report on the Ocean and Cryosphere, showing the impacts of climate change reflected in the speed of melting of ice sheets, mountain glaciers, and sea ice, and their implications for sea-level rise and marine ecosystems.

This new social movement and the very latest science inspired many corporations to take a leading role. Microsoft has set the agenda for the technology sector with the ambitious target to go carbon negative by 2030. By 2050, they want to remove all the carbon pollution from the atmosphere that they and their supply chain have emitted since the founding of the company in 1975. Sky has set the agenda for the media sector, pledging that they and their supply chain will go carbon negative by 2030. BP has also declared that it will be carbon neutral by 2050 by eliminating or offsetting over 415 million tons of carbon emissions. These companies form part of a group of over 1,000 global companies that have pledged to adopt science-based targets. Science-based targets effectively mean they will all be at net zero carbon emissions by 2050.

Given all this pressure, governments around the world in 2019 started to declare that we are, in fact, in a climate emergency, and that action has to be taken. At the time of publication, over 1,400 local governments and over 35 countries have made climate emergency declarations. Despite the fact that in 2020 the whole world was focused on dealing with the Covid-19 pandemic, climate change has remained a major issue. There have been lots of debates in the media and on social media about how the world could rebuild the post-Covid-19 economy in a more sustainable and low-carbon way. Many of the suggested ideas are discussed in Chapter 9 and many have already been implemented.

Chapter 3
Evidence for climate change

Science is not a belief system. It is a rational, logical methodology that moves forward by using detailed observation and experiments to constantly test and retest ideas and theories. It is the very foundation of our global society. So you cannot pick and choose which bits of scientific evidence you want to believe in and which bits you want to reject. For example, you cannot decide that you believe in antibiotics (as they may save your life) or that heavy metal tubes with wings can fly (because you want to go on holiday), and yet at the same time deny that smoking causes cancer, or that HIV causes AIDS, or that GHGs cause climate change. In this chapter, I present the scientific evidence that anthropogenic climate change is already happening.

Weight of evidence

If we are to understand climate change, we must understand how science works. The 'weight of evidence' principle prompts the constant need to compile new data and undertake new experiments in order to continually test our ideas and theories regarding climate. Over the past 40 years, the theory of climate change must have been one of the most comprehensively tested ideas in science. There are six main areas of evidence that should be considered:

1. We have tracked the rise in GHGs in the atmosphere and understand their role in past climate variations.

2. We know from laboratory and atmospheric measurements that GHGs do indeed absorb heat when they are present in the atmosphere. Table 1 summarizes the latest understanding of the main GHGs.

3. We have tracked significant changes in global temperatures and sea-level rise over the last century.

4. We have analysed the effects of natural changes on climate, including sunspots and volcanic eruptions, and though these are essential to understanding the pattern of temperature changes over the past 150 years, they cannot explain the warming trend (Figure 6).

5. We have observed significant changes in the Earth climate system, including the melting of the Greenland and Western Antarctic ice sheets; retreating of Arctic sea ice; retreating mountain glaciers on all continents; and shrinking permafrost and increased depth of its active layer (the top of the permafrost, which melts every summer).

6. We continually track global weather, and have seen significant shifts in the number and intensity of extreme events: climate change has now been demonstrated as a significant contributing factor to many of these extreme weather events.

In this chapter, we will consider evidence for changes in global temperature, precipitation, sea level, and extreme weather events.

Temperature

Temperatures can be estimated from a number of sources, both direct thermometer-based and proxy-based indicators. Proxy-based indicators are variables that are measured when direct measurements are not available or possible. For example, infrared

Table 1. Main greenhouse gases and their comparative ability to warm the atmosphere

Greenhouse gas	Chemical formula	Lifetime (years)	Pre-industrial levels	2018 levels	Human source	Global warming potential (compared to CO_2)	
						20 years	100 years
Carbon dioxide	CO_2		278 ppmv	407 ppm (>45% increase)	Fossil-fuel combustion Land-use changes Cement production	1	1
Methane	CH_4	12.4	700 ppbv	1,859 ppb (250% increase)	Fossil fuels Rice paddies Waste dumps Livestock	96	32
Nitrous oxide	N_2O	121	275 ppbv	331 ppb (>20% increase)	Fertilizer Industrial processes Fossil-fuel combustion	264	265

CFC-12	CCl_2F_2	100	Not naturally occurring	508 ppt	Liquid coolants / foams	10,800	10,200
HCFC-22	$CHClF_2$	11.9	Not naturally occurring	244 ppt	Liquid coolants	5,280	1,760
Perfluoro methane PCF-14	CF_4	50,000	0*	79 ppt	Production of aluminium	4,880	6,630
Sulphur hexa-fluoride	SF6	3,200	0*	9.59 ppt	Dielectric fluid	17,500	23,500

ppm = parts per million in the atmosphere
ppb = parts per billion in the atmosphere
ppt = parts per trillion in the atmosphere
* trace amounts are found naturally

(heat) satellite measurements are examples of a proxy that can be used to estimate surface temperatures.

Direct thermometer-based measurements of air temperature have been recorded at a number of sites in North America and Europe from as far back as 1760. The number of observation sites did not increase to a sufficient, worldwide geographical coverage to permit a global land average to be calculated until about the middle of the 19th century. Sea-surface temperatures (SSTs) and marine air temperatures (MATs) were systematically recorded by ships from the mid-19th century, but even today the coverage of the Southern Hemisphere is extremely poor. All these data sets require various corrections to account for changing conditions and measurement techniques. For example, for land data each station has been examined to ensure that conditions have not varied through time as a result of changes in the measurement site, instruments used, instrument shelters, or the way monthly averages were computed. We must also account for the growth of cities around some of the sites, which leads to warmer temperatures caused by the urban heat island effect. In the IPCC science report, the influence of the urban heat island effect is acknowledged as real and, if it were not corrected for, it would still be negligible for the global temperature compilation (less than 0.006°C).

For SST and MAT, there are a number of corrections that have to be applied. First, up to 1941 most SST temperature measurements were made in seawater hoisted on deck in a bucket. Since 1941, most measurements have been made at the ships' engine water intakes. Second, between 1856 and 1910 there was a shift from wooden to canvas buckets, which changes the amount of cooling caused by evaporation that occurs as the water is being hoisted on deck. In addition, through this period there was a gradual shift from the use of sailing ships to steamships, which altered the height of the ship decks and the speed of the ships, both of which can affect the evaporative cooling of the buckets. The other key correction that has to be made is for the global distribution of meteorological stations through time, which has varied greatly since 1870.

The collation of the global temperature records has been undertaken by a number of groups around the world, including the UK Meteorological Office, National Aeronautics and Space Administration (NASA), National Oceanic and Atmospheric Administration (NOAA), and the Japan Meteorological Agency (see Figure 8). In 2012, Professor Richard Muller, a physicist and previously a climate change sceptic, and his Berkeley group collated global temperature records for the last 250 years. Because his group had not taken account of all the corrections, their estimation of global warming was higher than that of the other groups. This was subsequently revised, and Muller publicly announced he had changed his mind and that climate change was occurring and was clearly due to human activity.

By making all the necessary corrections it is possible to produce a continuous record of global surface temperature from 1880 to 2020, which shows an observed warming of between 1.0°C and 1.3°C, with 1.1°C the most likely rise over this period (Figure 8). These observations are supported by 60 years of balloon and satellite data. For example, there are over 800 stations that twice a day release rawinsondes (meteorological instruments), or balloons, to measure temperature, relative humidity, and pressure through the atmosphere to a height of about 20 km, where they burst. The temperature records also show us that the land is warming up faster than the oceans. Since 1850, the land has warmed by 1.44°C and the oceans by 0.89°C (Figure 9).

Global temperatures have also been reconstructed for periods of time pre-dating instrumental or thermometer records. This has been achieved by using palaeoclimate proxies such as the thickness of tree rings and the isotopic composition of ice cores or cave deposits to estimate local temperatures. Combining the GMT instrumental record with the longer palaeoclimate temperature

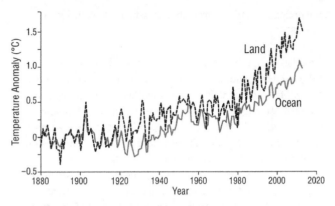

9. Land and ocean temperatures since 1850.

records shows a steep rise at the end of the record and is referred to as the global warming 'hockey stick'. A study, published in *Nature* in 2019, led by Raphael Neukom at the Oeschger Centre for Climate Change Research (University of Bern, Switzerland) used over 700 palaeoclimate records and showed that in the last 2,000 years, the only time the climate all around the world has

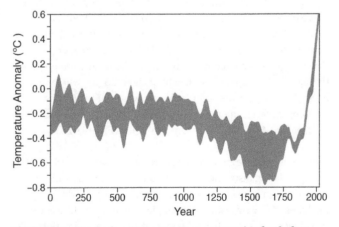

10. Northern Hemisphere temperature reconstruction for the last 2,000 years.

changed at the same time and in the same direction has been in the last 150 years, when over 98% of the surface of the planet has warmed (see Figure 10).

Precipitation

There are two global precipitation data sets: Hulme and the Global Historical Climate Network (GHCN). Unfortunately, unlike temperature, rainfall and snow data are not as well documented, and recording has not been carried out for as long. It is also known that precipitation over land tends to be underestimated by up to 10–15% owing to the effects of airflow around the collecting dish. Without correction of this effect, a spurious upward trend could be perceived in global precipitation. Despite these problems there seems to be a significant increase of precipitation over the past 25 years (see Figure 11), particularly in the Northern Hemisphere middle latitudes. This is supported by evidence that since the 1980s atmospheric water content has increased over the land and ocean as well as in the upper troposphere. This is consistent with the extra water vapour that the warmer atmosphere can hold.

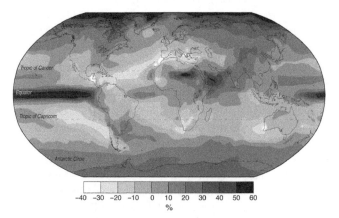

11. **Global precipitation changes (1900–2018).**

There is evidence for a global increase in precipitation but the evidence for this change is much stronger when considering individual regions. The latest IPCC report suggests that significant increases in precipitation have occurred in the eastern parts of North and South America, northern Europe, and northern and central Asia. It seems that seasonality of precipitation is also changing, for example in the high latitudes in the Northern Hemisphere, with increased rainfall in the winter and a decrease in the summer. Long-term drying trends have been observed on the Sahel, in the Mediterranean, southern Africa, and parts of southern Asia. It has also been observed that the amount of rain falling during heavy, 'extreme' rain events has increased.

Relative global sea level

The IPCC has also compiled all the current data on global sea level. They show that between 1901 and 2018, the global sea level rose by over 24 cm (see Figure 12). Sea-level change is difficult to measure, as relative sea-level changes have been derived from two very different data sets—tide-gauges and satellites. In the conventional tide-gauge system, the sea level is measured relative to a land-based tide-gauge benchmark. The major problem is that the land surface is much more dynamic than one would expect, with a lot of vertical movements, and these become incorporated into the measurements. Vertical movements can occur as a result of normal geological compaction of delta sediments, the withdrawal of groundwater from coastal aquifers, uplift associated with colliding tectonic plates (the most extreme of which is mountain-building such as in the Himalayas), or ongoing post-glacial rebound, and compensation elsewhere, associated with the end of the last ice age. The rebound is caused by the rapid removal of weight when the giant ice sheets melted, so that the land that has been weighed down slowly rises back to its original position. An example of this is Scotland, which is rising at a rate of 3 millimetres (mm) per year, while England is still sinking at a rate of 2 mm per year, despite the Scottish ice sheet having melted

10,000 years ago. In comparison, the problem with the satellite data is that the time covered is too short. The best satellite data started in January 1993 and show a trend of over 35 mm rise in sea level per decade. This means satellite data have to be combined with the tide-gauge data to look at long-term trends.

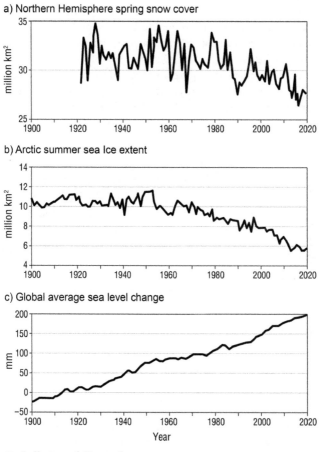

a) Northern Hemisphere spring snow cover

b) Arctic summer sea Ice extent

c) Global average sea level change

Year

12. **Indicators of climate change.**

In summary, between 1901 and 2018 the global average sea level rose by about 2 mm per year; with the fastest rise in sea level observed between 2008 and 2018 at 4.2 mm per year. The sea-level rise of the last 30 years is made up of the following contributions: 39% from thermal expansion of the ocean; 9% Antarctic ice sheet; ~12% Greenland ice sheet; 27% glaciers and other ice caps; and another ~13% due to the overall reduced land storage of water (Figure 13). The Greenland and Antarctic ice sheets have contributed to recent sea-level rise, and this contribution is accelerating. At the moment it is estimated that Greenland is losing over 230 gigatonnes (Gt) of ice per year, a seven-fold increase since the early 1990s. Meanwhile, Antarctica is losing about 150 Gt of ice per year, a five-fold increase since the early 1990s, and most of this loss is from the northern Antarctic peninsula and the Amundsen sea sector of west Antarctica.

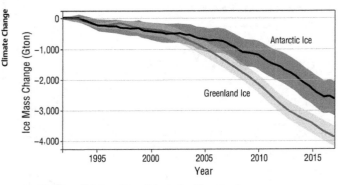

13. Melting of Antarctic and Greenland ice sheets.

Other evidence for global warming

Other evidence for climate change comes from the high latitudes and from monitoring extreme weather events. The annual mean Arctic sea ice extent has decreased in total between 1979 and 2018 at a rate of 3.5–4.1% per decade, which means a loss of between 0.45 and 0.51 million km^2 per decade. The summer sea ice minimum has decreased even more by 12.8% per decade, which is equivalent to a loss of 1 million km² per decade. In contrast, between 1979 and 2018 the annual mean Antarctic sea ice extent has varied markedly with record highs and lows but there is no significant trend when the continuous satellite observations for the period are examined.

There is also evidence from permafrost regions. Permafrost exists in high-latitude and high-altitude areas, where it is so cold that the ground is frozen solid to a great depth. During the summer months, only the top half-metre or so of the permafrost becomes warm enough to melt, and this is called the 'active layer'. There has been a 3°C warming in Alaska and 2°C warming in northern European/Russian permafrost over the last 50 years, and evidence that this active layer has become much deeper. The maximum area covered by seasonal permafrost has decreased by 7% in the Northern Hemisphere since 1900, with a decrease in the spring of up to 15%. This increasingly dynamic cryosphere will amplify the natural hazards for people, structures, and communication links. Already we have seen this in the form of damage to buildings, roads, and pipelines, such as to the oil pipelines in Alaska. In addition, there is evidence that most if not all non-ice-sheet glaciers are in retreat. The amount of total snowfall and the annual snow and ice cover, particularly in the Northern Hemisphere, has greatly reduced (Figure 12). Between 1922 and 2018, over 0.27 million km² of snow and ice cover per decade has been lost. In the Arctic, snow-cover duration has decreased on average by ~3–5 days per decade and larger declines have occurred in the Eurasian Arctic region (~12.6 days) and North American Arctic region (6.2 days).

There is also evidence that spring is occurring earlier in the Northern Hemisphere. The ice cover records from the Tornio River in Finland, which have been compiled since 1693, show that the spring thaw of the frozen river now occurs a month earlier. In Kyoto, Japan, the famous cherry blossoms now appear 21 days earlier than 100 years ago. In France, the grape harvest in Beaune is now 10 days earlier than 100 years ago. In Britain, among the variety of indicators of an earlier spring is evidence of birds nesting over 12 days earlier than 45 years ago. Insect species—including bees and termites—that need warm weather to survive are moving northward, and some have already reached England by crossing the Channel from France. Meanwhile, in the USA, species that are active in early spring such as lilac and honeysuckle are unfurling their leaves 3 weeks earlier than 40 years ago.

Extreme weather events

The latest IPCC report states that it is virtually certain that anthropogenic climate change has caused increases in the frequency and severity of hot extremes and decreases in cold extremes on most continents. The frequency and intensity of heat waves has increased in Europe, Asia, America, and Australia. The past decade has seen record-breaking heatwaves occurring in Australia, Canada, Chile, China, India, Japan, the Middle East, Pakistan, and the USA.

Climate change is also the main cause of the intensification of heavy precipitation observed over continental regions, often resulting in flooding. Record-breaking extreme floods have been recorded over the past decade in Brazil, Britain, Canada, Chile, China, East Africa, Europe, India, Indonesia, Japan, Korea, the Middle East, Nigeria, Pakistan, South Africa, Thailand, the USA, and Vietnam.

Human climate change has also played a role in shaping the global distribution and intensity of tropical cyclones. A 2020 study by James P. Kossin at NOAA and colleagues showed a 15% increase in the occurrence of the most destructive cyclones around the world over the past 40 years. Most marked was the 49% per decade increase in major hurricanes occurring in the North Atlantic and the 18% per decade increase in major cyclones in the southern Indian Ocean. In summary, the number of tropical cyclones originating in the North Atlantic Ocean, Pacific Ocean, and southern Indian Ocean has increased as well as the year-to-year variability. For example, in 2019 there were four spectacular cyclones in the Indian Ocean and two of these, in the southern Indian Ocean, were unprecedented. The 2018–19 south-west Indian Ocean cyclone season was the costliest and most active season ever recorded since reliable records began in 1967. In 2020, Super Cyclonic Storm Amphan formed in the North Indian Ocean and made landfall in west Bengal, affecting nearly 40 million people and causing over $13 billion of damage.

The reason why scientists are certain that many of these extreme weather events are made worse by climate change is the new field of attribution science. Advances in computer processing power and improved methods for modelling the factors that contribute to weather allow scientists to run weather simulations for a region with and without anthropogenic GHG forcing. This allows us to determine the extent to which climate change has contributed to specific extreme weather events and, if there has been a contribution, whether it has increased the intensity or the frequency or both. Over 113 extreme weather events that occurred between 2015 and 2020 have been studied using attribution science: 70% of events were found to have increased frequency or intensity due to climate change; 26% were found to have a reduced occurrence due to climate change; and 4% showed no variation due to climate change.

What do the climate change deniers say?

One of the best ways to summarize the evidence for climate change is to review what the climate change deniers say against the current state-of-the-art science.

Ice-core data suggest atmospheric CO_2 responds to global temperature, therefore atmospheric CO_2 cannot cause global *temperature changes*. At the end of the last ice age the Earth warmed up. We know from Greenland and Antarctica ice cores that the Northern and Southern Hemispheres warmed up at different times and at different rates. On top of this there are millennial-scale climate events, when huge amounts of ice broke off from the melting North American ice sheet, flooding the North Atlantic Ocean with freshwater, changing ocean circulation, and cooling the Northern Hemisphere.

One of these events, called Heinrich event 1, occurred about 15,000 years ago, and the other was the Younger Dryas, which occurred about 12,000 years ago. Because of the wonderfully named 'bipolar climate seesaw', whenever the Northern Hemisphere cools down heat is exported southwards by the oceans and the Southern Hemisphere warms up. So if you compare an individual ice-core temperature record with reconstructed atmospheric CO_2 levels then there will be times when the relationship seems to swap. To really understand the relationship between global temperatures and CO_2, Professor Jeremy Shakun of Harvard University and colleagues created a master stack of all the temperature records across the end of the last ice age (see Figure 14). This shows that atmospheric CO_2 level leads global temperatures, adding to our confidence that it was contributing to the warming of the Earth as we exited the last great ice age.

CO_2 is a small part of the atmosphere—it can't have a large heating effect. This is an attempt to play a classic common-sense argument, but it is completely wrong. First, scientists have repeated experiments in the laboratory and taken measurements in the atmosphere, demonstrating again and again the greenhouse effect of CO_2. Second, as for the 'common-sense'-scale argument that a very small quantity of something can't have much of an effect, it only takes 0.1 grams of cyanide to kill an adult, which is about 0.0001% of your body weight. Compare this with CO_2,

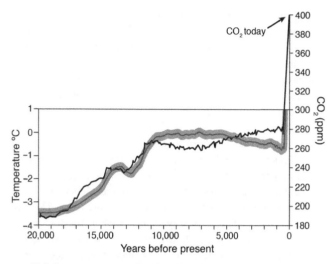

14. Global temperatures and carbon dioxide changes for the last 20,000 years.

which currently makes up 0.04% of the atmosphere and is a strong GHG. Meanwhile, nitrogen makes up 78% of the atmosphere and yet it is highly unreactive.

Every data set has been corrected or tweaked to show global warming. For people who are not regularly involved in science, this seems to be the biggest problem with the whole 'climate change has happened' argument. As shown above, all the climate data sets covering the last 150 years require some sort of adjustment. This, though, is part of the scientific process. For example, in 2012, Muller and his Berkeley group published their collated global temperature records and showed an increase of 1.5°C over the past 250 years. This was much higher than other estimates, as the Berkeley group had not corrected all the climate records. Science moves forward incrementally; it gains more and more understanding and insights into the data sets it is using.

This constant questioning of all data and their interpretations is the core strength of science: each new correction or adjustment is due to a greater understanding of the data and the climate system, and thus each new study adds to the confidence that we have in the results. This is why the IPCC report refers to the 'weight of the evidence', since our confidence in science increases if similar results are obtained from very different sources.

Recent changes in global temperatures are due to changes in the Sun. Both deniers and climate scientists agree that sunspots and volcanic activity do influence climate and global temperatures. The difference between the two camps is that deniers want these natural variations to be the dominant control on climate. There is evidence that the 11-year solar cycle, during which the Sun's energy output varies by roughly 0.1%, can influence ozone concentrations, temperatures, and winds in the stratosphere. These changes have only a very small effect on surface temperatures. Figure 15 shows that since 1880, the solar radiation increased gradually to a peak in about 1955, and since then it has been decreasing. So over the past 50 years, when global temperatures have increased dramatically, solar output has in fact decreased.

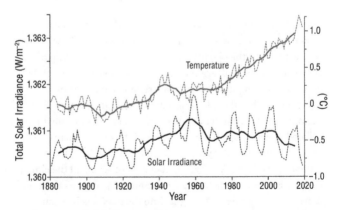

15. Sunspot and global temperatures.

Over the past 150 years, significant changes in climate have been recorded, changes which are markedly different from those of at least the past 2,000 years. These changes include a 1.1°C increase in average global temperatures; sea-level rise of over 24 cm; significant shifts in the seasonality and intensities of precipitation; changing weather patterns; the accelerated melting of the Greenland and Western Antarctic ice sheets; and the significant retreat of Arctic sea ice and nearly all continental glaciers. According to the US National Oceanic and Atmospheric Administration, between 1880 and 2020, the 10 warmest years on record have all occurred within the past 15 years, with 2020 joint warmest year with 2016, followed by 2019, 2015, 2017, 2018, 2014, 2010, 2013, and 2005. The IPCC 2021 report states that the evidence for climate change is unequivocal, and there is very high confidence that this warming is due to human emissions of GHGs. This statement is supported by six main lines of evidence: (1) the rise in GHGs in the atmosphere has been measured and the isotopic composition of the gases shows that the majority of the additional carbon comes from the burning of fossil fuels; (2) laboratory and atmospheric measurements show that these gases absorb heat; (3) significant changes in global temperatures and sea-level rise have been observed over the past century; (4) other significant changes have been observed in the cryosphere, oceans, land, and atmosphere including retreating ice sheets, sea ice, and glaciers, and extreme weather events, all of which can be directly attributed to the impact of climate change; (5) there is clear evidence that natural processes including sunspots and volcanic eruptions cannot explain the warming trend over the past 100 years; and (6) we now have a deeper understanding of the longer term climate changes of the past and the critical role GHGs have played in regulating the climate of our planet.

Chapter 4
Modelling future climate

The whole of human society operates on knowing the future weather. For example, farmers in India know when the monsoon rains will come next year and so they know when to plant the crops. Farmers in Indonesia know there are two monsoon rains each year, so next year they can have two harvests. This is based on their knowledge of the past, as the monsoons have always come at about the same time each year in living memory. But the need to predict goes deeper than this; it influences every part of our lives. Our houses, roads, railways, airports, offices, cars, trains, and so on are all designed for the local climate. For example, in England all the houses have central heating, as the outside temperature is usually below 20°C, but no air-conditioning, as temperatures rarely exceed 26°C, while in Australia the opposite is true: most houses have air-conditioning but rarely central heating. Predicting future climate is now essential, as we can no longer rely on records of past weather to tell us what the future will hold. We also need to understand the consequences of our actions. For example, if we continue to emit GHGs at the same rate as today, how much climate change will occur? So we have to develop new ways of understanding potential futures. We model the future (Figure 16).

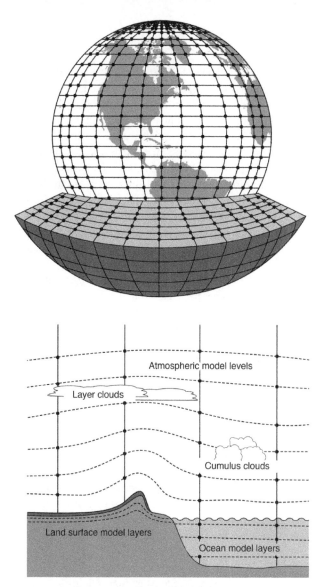

16. **Generic structure of a global climate model.**

Models

There is a whole hierarchy of climate models, from relatively simple box models to extremely complex three-dimensional GCMs. Each has a role in examining and furthering our understanding of the global climate system. It is the complex three-dimensional GCMs that are used to predict future global climate. These comprehensive climate models are based on physical laws represented by mathematical equations, which are solved using a three-dimensional grid over the globe. To obtain the most realistic simulations, all the major parts of the climate system must be represented in sub-models, including atmosphere, ocean, land surface (topography), cryosphere, and biosphere, as well as the processes that go on within them and between them.

Over the past 40 years there has been a huge improvement in climate models. This has been due to our increased knowledge of the climate system but also because of the nearly exponential growth in computer power. There has been a massive improvement in spatial resolution of the models from the very first IPCC report in 1990 to the latest in 2021. The current generation of GCMs have multiple layers in the atmosphere, land, and ocean and can have a spatial resolution greater than one point every 30 km by 30 km. Equations are typically solved for every simulated 'half-hour' of a model run. Many physical processes, such as atmospheric chemistry, the formation of clouds, production and movement of aerosols (particles suspended in the air), and ocean convection, take place on a much smaller scale than the main model can resolve. The effects of small-scale processes have to be lumped together, which is referred to as 'parameterization'. All of these parameterizations are checked with separate 'small-scale-process models' to validate the scaling up of these smaller influences.

The biggest unknown in the models is not the physics or the chemistry or the biology: it is the estimation of future global GHG emissions over the next 80 years. This includes many variables, from the global economy to personal lifestyles. Individual models are therefore run many times with different emission scenarios to provide a range of changes that could occur in the future. In fact, the latest (sixth) IPCC Assessment Report (AR6) has compiled the results of multiple runs from over a hundred distinct climate models being produced across forty-nine different international modelling groups, which are all part of the latest (sixth) Coupled Model Intercomparison Projects (CMIP6). Of course, as computer processing power continues to increase, both the representation of coupled climate systems and the spatial scale will continue to improve.

Carbon cycle

At the heart of the climate models is the carbon cycle, central to estimating what happens to anthropogenic CO_2 and CH_4 emissions. The Earth's carbon cycle is complicated, with both large sources and sinks of CO_2. Currently half of all our carbon emissions are absorbed by the natural carbon cycle and do not end up in the atmosphere but rather in the oceans and the terrestrial biosphere. Figure 17 shows the global reservoirs in gigatonnes of carbon (GtC; or 1,000 million tonnes) and fluxes (the ins and outs of carbon in GtC per year). These indicative figures show the changes since the Industrial Revolution. Evidence is accumulating that many of the fluxes can vary significantly from year to year.

This is because in contrast to the static view conveyed in figures like this one, the carbon system is dynamic, and coupled to the climate system on seasonal, interannual, and decadal timescales. What has become clear is that the ocean surface and the land biosphere each take up about 25% of our carbon emissions every year. As the oceans continue to warm they can hold less dissolved CO_2, which means their uptake will reduce. As we continue to deforest and substantially alter land use, the ability of the land biosphere to absorb carbon diminishes.

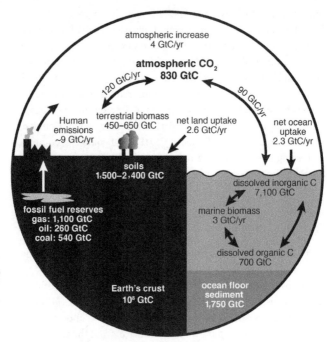

17. **The carbon cycle, in gigatonnes of carbon.**

Warming and cooling effects

As well as the warming effects of the GHGs, the Earth's climate
system is complicated in that there are also cooling effects (see
Figure 18). These include the amount of aerosols in the air (many
of which come from human pollution, such as sulfur emissions
from power stations), which have a direct effect on the proportion
of solar radiation that hits the Earth's surface. Aerosols have a
significant local or regional impact on temperature. Computer
simulations of climate change demonstrate that industrial areas of
the planet have not warmed as much as would be predicted just
from rising GHGs. This so called 'global dimming', or more

precisely 'regional dimming', has been confirmed with real temperature and aerosol measurements. Water vapour is a GHG, but, at the same time, the upper white surface of clouds reflects solar radiation back into space. The level of reflectivity of a surface is called 'albedo'. Clouds and ice have a high albedo, which means that they reflect large quantities of solar radiation away from surfaces on Earth. Increasing aerosols in the atmosphere increases the amount of cloud cover, as they provide points on which the water vapour can nucleate. Predicting what will happen to the amount and types of clouds, and their warming or cooling potential, has been one of the key challenges for climate scientists.

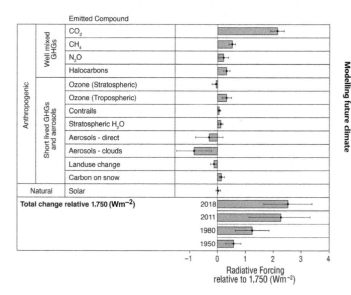

18. **Radiative forcings between 1750 and 2018.**

Emission models of the future

As noted earlier, a critical problem with trying to predict future climate is predicting the amount of CO_2 emissions that will be produced in the future. This will be influenced by population growth, economic growth, development, fossil-fuel usage, the rate at which we switch to alternative energy, the rate of deforestation, and the effectiveness of international agreements to cut emissions. Out of all the systems that we are trying to model into the future, humanity is by far the most complicated and unpredictable. If you want to understand the problem of predicting what will happen in the next 80 years, imagine yourself in 1920 and what you would have predicted the world to be like in the 21st century. At the beginning of the 20th century, the British Empire was the dominant world power due to the Industrial Revolution and the use of coal. Would you have predicted the switch to a global economy based on oil after the Second World War? Or the explosion of car use? Or the general availability of air travel? Even 30 years ago, it would have been difficult to predict that there would be budget airlines, allowing for cheap flights throughout Europe, the USA, and Asia.

The first IPCC reports used simplistic assumptions of GHG emissions over the next 100 years. From 2000 onwards, the climate models used more detailed scenarios set down in an IPCC special report (Special Report on Emission Scenarios by the IPCC, or SRES, 2000). The 2013 IPCC AR5 used more sophisticated representative concentration pathways (RCPs), which considered a much wider variable input to the socioeconomic models, including population, land use, energy intensity, energy use, and regional differentiated development. The RCPs were defined by the final radiative forcing achieved by the year 2100, and they range from 2.6 to 8.5 watts per square metre (W/m^2). Radiative forcing is defined as the difference between sunlight (radiant energy) received by the Earth and the energy radiated back to space and is measured in units of W/m^2 of the Earth's surface. For the 2021 IPCC AR6, the RCP1.9 was added, to represent the Paris 2015 agreement ambition to keep global temperature rise to only 1.5°C above pre-industrial levels.

The IPCC AR6 also uses shared socioeconomic pathways (SSPs), which were developed to cover the full range of possible futures. The SSPs are a set of five narratives and driving forces that may shape the global economy and global emissions in the future (Figure 19). The SSPs were defined in 2017 by Keywan Riahi and colleagues in a paper in the journal *Global Environmental Change* as described below.

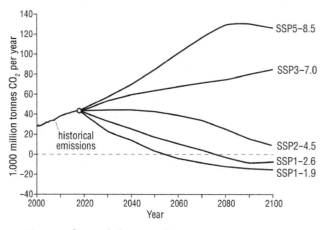

19. Future carbon-emission scenarios.

- SSP1: Sustainability—taking the green road

Low challenges to mitigation and adaptation: In this scenario the world shifts gradually but continually towards a more sustainable path, with inclusive and environmentally aware economic development. Management of the global commons slowly improves, educational and health investments accelerate the demographic transition towards reducing population, and the emphasis on economic growth shifts towards a broader emphasis on human wellbeing. Driven by an increasing commitment to achieving development goals, inequality is reduced both across and within countries. Consumption is oriented towards low material growth and lower resource and energy intensity.

- SSP2: Middle of the road

Medium challenges to mitigation and adaptation: In this scenario the world follows a path in which social, economic, and technological trends do not shift markedly from historical patterns. Development and income growth proceeds unevenly, with some countries making relatively good progress while others fall short of expectations. Global and national institutions work towards but make slow progress in achieving sustainable development goals. Environmental systems experience degradation, although there are some improvements and overall the intensity of resource and energy use declines. Global population growth is moderate and levels off in the second half of the century. Income inequality persists or improves only slowly, and challenges to reducing vulnerability to societal and environmental changes remain.

- SSP3: Regional rivalry—a rocky road

High challenges to mitigation and adaptation: In this scenario there is a resurgence of nationalism, concerns about competitiveness and security, and regional conflicts that push countries to increasingly focus on domestic or, at best, regional issues. Policies shift over time to become increasingly oriented towards national and regional security issues. Countries focus on achieving energy and food security goals within their own regions at the expense of broader based development. The scenario also assumes that investments in education and technological development decline. Economic development is slow, consumption is material-intensive, and inequalities persist or worsen over time. Population growth is low in industrialized and high in developing countries. A low international priority for addressing environmental concerns leads to strong environmental degradation in some regions.

- SSP4: Inequality—a road divided

Low challenges to mitigation, but high challenges to adaptation: In this scenario there are highly unequal investments in human capital, combined with increasing disparities in economic opportunity and political power, lead to increasing inequalities and stratification both across and within countries. Over time, a gap widens between an internationally connected society that contributes to knowledge- and capital-intensive sectors of the global economy, and a fragmented collection of lower income, poorly educated societies that work in a labour-intensive, low-tech economy. Social cohesion degrades, and conflict and unrest become increasingly common. Development is substantial in the high-tech economy. The globally connected energy sector diversifies, with investments in both carbon-intensive fuels like coal and unconventional oil, but also low-carbon energy sources. Environmental policies focus on local issues around middle- and high-income areas.

- SSP5: Fossil-fuelled development—taking the highway

High challenges to mitigation, but low challenges to adaptation: In this scenario the world places increasing faith in competitive markets, innovation, and participatory societies to produce rapid technological progress and development of human capital as the path to sustainable development. Global markets are increasingly integrated. There are also strong investments in health, education, and institutions to enhance human and social capital. At the same time, the push for economic and social development is coupled with the exploitation of abundant fossil-fuel resources and the adoption of resource- and energy-intensive lifestyles around the world. All these factors lead to rapid growth of the global economy, while global population peaks and declines in the 21st century. Local environmental problems like air pollution are successfully managed. There is faith in the ability to effectively manage social and ecological systems, including by geoengineering if necessary.

These are narratives that describe different pathways our future society could take. SSP1 and SSP5 are optimistic about human development, with both allowing for 'substantial investments in education and health, rapid economic growth, and well-functioning institutions'. The difference is that SSP5 is fossil-fuel energy intensive, while SSP1 assumes a shift to renewable energy. SSP3 and SSP4 are pessimistic about the future and SSP2, as it says, is a middle-of-the-road scenario. The scenarios used in the AR6 are a combination of the SSPs and RCPs, providing a clear narrative and an outcome. This is because the SSPs do not include any mitigation measures, so a high-emission global economy could have a lower concentration pathway by employing huge amounts of mitigation. If all the SSP and RCP combinations are examined, then some are extremely unlikely and others are almost impossible, for example SSP5 and RCP1.9. The AR6 focuses on five main scenarios SSP1–1.9, SSP1–2.6, SSP2–4.5, SSP3–7.0, and SSP5–8.5 (Table 2).

Modelling uncertainty

In the most recent IPCC AR6, the past- and future-emission scenarios were used in about 100 distinct, independent GCMs. Each of these models has its own independent design and parameterizations of key processes. The independence of each model is important, as confidence may be derived from multiple runs on different models providing similar future climate predictions. In addition, the differences between the models can help us to learn about their individual limitations and advantages. Within the IPCC, due to political expediency, each model and its output is assumed to be equally valid. This is despite the fact that

Table 2. Defining representative concentration pathway

Representative concentration pathway (RCP)	Description
RCP8.5	Rising radiative forcing pathway leading to 8.5 W/m² (~1,000 ppm CO_2 between 2081 and 2100)
RCP7	Stabilization pathway to 7 W/m² (~800 ppm CO_2 between 2081 and 2100)
RCP4.5	Stabilization pathway to 4.5 W/m² (~600 ppm CO_2 between 2081 and 2100)
RCP2.6 (also called RCP3PD)	Peak in radiative forcing at ~3 W/m² (peak of ~490 ppm CO_2 and then negative emission to ~450 ppm CO_2 between 2081 and 2100 leading to 2.6 W/m² by 2100)
RCP1.9 (also called the 1.5°C scenario)	Peak in radiative forcing at ~1.9 W/m² (~400 ppm CO_2 between 2081 and 2100; all the SSPs require negative global emissions to achieve this pathway)

some are known to perform better than others when tested against the reality provided by the historical and palaeoclimate records. Moreover, though we understand uncertainty within a single model, the notion of quantifying uncertainty from many models currently lacks any real theoretical background or basis. The IPCC combines all the models used for each run and then presents the mean and the uncertainty between the models. This way it is clear that there are differences in the models' output but that in general they agree and show very different futures based on which scenario we take. The uncertainties in the IPCC 2021 report are slightly higher than those in previous reports. This is because of our greater understanding of the processes and our ability to quantify the uncertainty of our knowledge. So although our confidence in the climate models has increased, so has the range of possible answers for any specific GHG forcing. One way to test the models and their uncertainty is to compare their predictions with the real world outcome. The CMIP3 took place just before the IPCC AR3 in 2001, and we can use those model predictions to compare with the following 20 years of real data. As can be seen from Figure 20 the predicted warming of the world by the earlier climate models was very good, and we have now had over 20 years to improve the models and expand the number we use in our predictions.

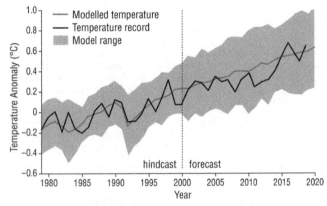

20. Climate model predictions compared with climate data (2000–20).

Future global temperatures, precipitation, sea ice, and sea level

Between ten and thirty-six climate models have been run for each of the five SSP-RCPs for the IPCC 2021 report to produce scenarios of global temperature, precipitation, sea-ice and sea-level changes that may occur by 2100. These climate models suggest that depending on our GHG emissions the global surface temperature could, by 2081–2100, rise between 1.3°C and 5.5°C compared with the pre-industrial period (1850–1900): see Table 3. In all the scenarios except SSP1–1.9, a global temperature rise of over 1.5°C is reached between 2021 and 2041 with the best

Table 3. Temperature, precipitation, and sea-level projections

Shared socioeconomic pathways	2081–2100 global temperature change relative to the pre-industrial period (°C)	2081–2100 global sea-level rise relative to 1990–2014 (m)	2081–2100 global precipitation rise relative to 1990–2014 (%)
SSP5–8.5	4.7 (3.4–5.5)	0.73 (0.50–1.07)	8.2% (2.5–13.8%)
SSP3–7.0	3.9 (2.8–4.6)	0.65 (0.41–1.00)	5.5% (0.5–10.4%)
SSP2–4.5	2.9 (2.1–3.3)	0.55 (0.28–0.89)	4.7% (1.7–6.4%)
SSP1–2.6	1.9 (1.4–2.2)	0.41 (0.29–0.71)	3.2% (0.7–5.6%)
SSP1–1.9	1.5 (1.1–1.7)	Not yet completed	2.7% (0.6–4.8%)

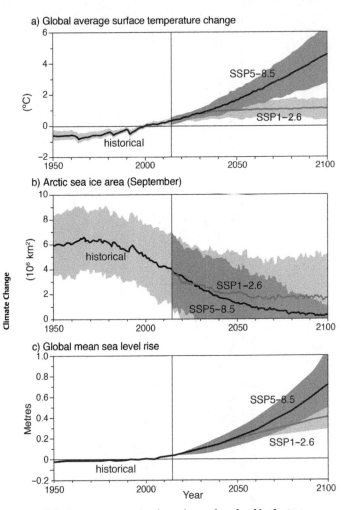

a) Global average surface temperature change

b) Arctic sea ice area (September)

c) Global mean sea level rise

21. Global temperatures, Arctic sea ice, and sea level in the 21st century.

estimate being 2030. The models also show that the temperature rise will be unevenly distributed, with the largest rises in temperature observed on land.

Future sea-level rise is dependent on which SSP we follow and could be between 0.32 m and 0.82 m in the last two decades of the century (Table 3 and Figure 21). With the 20 cm rise which has already occurred, this would represent a total rise of 0.52 m and 1.02 m. If we look at the final projected sea level at 2100, the models show an increase in global mean sea level of between 27 cm and 98 cm. This is similar but more extreme than the projection made by the IPCC 2007 report, which suggests a sea-level rise of between 28 cm and 79 cm by 2100.

Both average land and ocean precipitation are very likely to increase under all five of the SSPs (see Table 3). The annual global average land precipitation by 2081–2100 relative to 1995–2014 will increase by 2.7% (with a range of 0.6–4.8%) in the low-emission scenario SSP1–1.9 and 8.2% (with a range of 2.5–13.8%) in the high-emission scenario SSP5–8.5. Based on all the scenarios the average land precipitation will increase approximately 1–3% per 1°C of global warming.

In the three worst-case SSP scenarios (SSP2–4.5, SSP3–7.0, and SSP5–8.5), the Arctic Ocean will become effectively ice-free (coverage below 1 million km^2) in September (the minimum ice month) by 2081–2100.

What do the climate change deniers say?

One of the best ways to summarize the perceived problems of modelling climate change is to review what the climate change deniers say.

Different models give different results, so how can we trust any of them? This is a frequent response from many people not familiar with modelling, as there is a feeling that somehow science must be able to predict an exact future. However, in no other walk of life do we expect this precision. For example, you would never expect to get a perfect prediction on which horse will win a race or which football team in a match will emerge triumphant. The truth is that none of the climate models is right, because they provide a range of potential futures. Our view of the future is strengthened by the use of more than one model, because each model has been developed by different groups of scientists around the world, using different assumptions, different computers, and different programming languages; thus they provide their own independent future predictions. What causes scientists to have confidence in the model results is that they all predict the same trend in global temperature and sea level for the next 80 years. One of the great strengths of the 2021 IPCC reports is that they used over a hundred distinct models from forty-nine different modelling groups around the world compared to forty models in 2013, twenty-three models in 2007, and seven in 2001.

Climate models are too sensitive to CO_2. To dismiss the importance of climate change, many deniers argue that the models are oversensitive to changes in GHGs. This is a classic 'it is not as bad as you think' argument. The strength of having so many climate models is that scientists can also give an estimation of how confident they are in the model results and check how sensitive their models are compared with each other and real world data. One key test of climate models is the equilibrium climate sensitivity (ECS), whereby the model predicts what the global temperature change would be if pre-industrial CO_2 levels were doubled. These results have been very consistent over the past 50 years (see Figure 22), and the 2021 IPCC report suggests the most up to date modelled range is between 2.50°C and 5.43°C (average 3.74°C), which is consistent with other measures.

Climate models fail to reconstruct natural variability. Many climate deniers argue that the current warming trend is due to

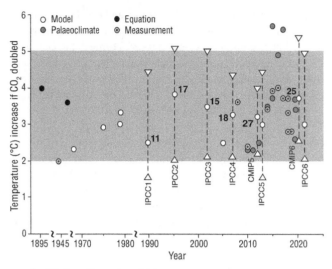

22. Equilibrium climate sensitivity.

natural variations. But scientists include all the natural variables, including ones that cool the climate, in the climate models. Combining all our scientific knowledge of natural (solar, volcanic, aerosol, and ozone) and human-made (GHGs and land-use change) factors warming and cooling the climate shows that 100% of the warming observed over the past 150 years is due to humans.

Clouds have negative feedbacks on global climate that will reduce the effects of climate change. As has been the case since the very first IPCC report in 1990, one of the uncertainties in the models is the role of the clouds and their interaction with radiation. Clouds can both absorb and reflect radiation, thereby cooling the surface, and absorb and emit long-wave radiation, warming the surface. The competition between these effects depends on a number of factors: the height, thickness, and radiative properties of clouds. The radiative properties and formation and development of clouds depend on the distribution of atmospheric water vapour, water drops, ice particles, atmospheric aerosols, and cloud thickness. The physical basis of how clouds are represented or parametrized in the climate models has greatly improved through the inclusion of representations of cloud microphysical properties in the cloud water budget equations. Figure 18 shows that even if the most extreme cooling value is applied for clouds, the warming factors due to GHGs are still three times larger.

Climate change must be caused by galactic cosmic rays (GCRs). GCRs are high-energy radiation that originates outside our solar system and may even be from distant galaxies. It has been suggested that they may help to seed or 'make' clouds. So reduced GCRs hitting the Earth would mean fewer clouds, which would reflect less sunlight back into space and would cause the Earth to warm up.

But there are two problems with this idea. First, the scientific evidence shows that GCRs are not very effective at seeding clouds. Second, over the past 50 years, the flux of GCRs has actually increased, hitting record levels in recent years. If this idea were correct, GCRs should be cooling the Earth, which they aren't.

Modelling future climate change is about understanding the fundamental physical processes of the climate system. Five new emission scenarios were produced for the 2021 IPCC science report, using a much wider set of inputs to the socioeconomic models, including population, land use, energy intensity, energy use, and regional differentiated development. One of these emissions pathways (SSP1-1.9) was developed to indicate to policy makers how they could achieve the aspired target of just 1.5°C warming set at the Paris 2015 climate change conference. Over a hundred climate models were used in developing the IPCC scenarios, providing a huge 'weight of evidence'. Using the three main realistic carbon-emission pathways over the next 80 years, the climate models suggest the global mean surface temperature could rise by between 2.1°C and 5.5°C by 2100. However, it must

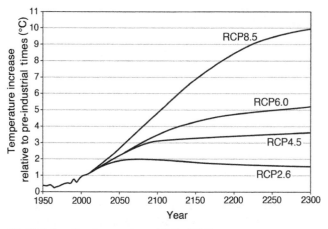

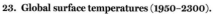

23. **Global surface temperatures (1950–2300).**

be remembered that global temperatures will not stop changing once we get to the year 2100. Figure 23 shows how temperatures could continue to rise way beyond the levels of this century, depending on the chosen emission pathway. Using the three main realistic carbon-emission pathways, the models also predict an increase in global mean sea level of between 0.50 m and 1.3 m by 2100 compared with pre-industrial times.

Chapter 5
Climate change impacts

This chapter assesses the potential impacts of climate change and how these alter in scale and intensity with increasing warming. The IPCC 'Impacts, Adaptation and Vulnerability' reports look at potential impacts on a regional level as well as by different sectors, such as freshwater resources, ecosystems, coastal and ocean systems, food security, and human health. It is also necessary to estimate the extent and magnitude of climate change impacts at national and local levels. There are a number of excellent national reports and tools, such as the US National Climate Assessment and the UK Climate Impacts Programme, both of which have interactive tools to understand the potential effects of climate change within their own countries. In this chapter, the potential impacts are broken down into sectors: extreme heat and droughts, storms and floods, agriculture, ocean acidification, biodiversity, and human health.

What is dangerous climate change?

One of the most important questions for policy makers is what is dangerous climate change? Because if we are to cut global GHG emissions we need a realistic target concerning the degree of climate change with which we can cope. In February 2005, the British government convened an international science meeting at Exeter, UK, to discuss this very topic. This was a very political science meeting, as the UK government was looking for a recommendation to take to the Group of Eight (G8) meeting in Gleneagles. At that time Britain held both the chair of the G8 and the presidency of the EU, and the then Prime Minister Tony Blair wished to push forward internationally his joint agenda of climate change reduction and poverty alleviation in Africa. The meeting and a lot of supporting research at the time suggested a limit of 2°C above pre-industrial average temperature: below this threshold there seemed to be both winners and losers due to regional climate change, but above this temperature everyone seemed to lose. It has now been shown that due to the impacts of extreme weather events there are in fact no regions that benefit from a 2°C warming. At the Paris climate change negotiation meeting in 2015 the Alliance of Small Island States (AOSIS) and some key developing countries reiterated that even a small amount of warming would be devastating for their countries. The Paris agreement set 2°C as the key target but added the aspiration of 1.5°C. Subsequently the IPCC special report on 1.5°C global warming published in 2018 supported this lower target by demonstrating that there is a significant increase in regional and national climate change impacts between a 1.5°C and 2.0°C world.

Extreme events and society's coping range

The single biggest problem with climate change is our inability to predict its effects on the future. Humanity can live, survive, and even flourish in extreme climates from the Arctic to the Sahara, but problems arise when the predictable extremes of local climate are exceeded. For example, heatwaves, storms, droughts, and floods in one region may be considered fairly normal weather in another. This is because each society has a coping range—a range of weather with which it can deal. Figure 24 shows the theoretical effect of combining the societal coping range with climate change. In our present climate, the coping range encompasses nearly all the variations in weather with maybe only one or two extreme events. These could be 1-in-200-year events that surpass the coping ability of that society. As the climate moves gently to its new average, if the coping range stays the same then more extreme events will occur. Hence a 1-in-200-year event may become a 1-in-50-year event. The good news is that the societal coping range is flexible and can adapt to a shifting baseline and more frequent extreme events—as long as there is strong climate science to provide clear guidance on what sort of changes are going to occur. The speed with which the societal coping range can expand depends on what aspect of society is being affected: adaptation of the individual's behaviour can be extremely quick, while building major infrastructure can take decades to complete. One of the biggest challenges of climate change is to build flexible and resilient societies that are able to adapt to a changing future.

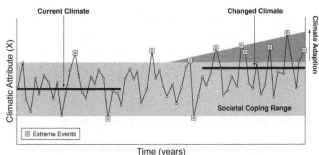

24. **Climate change, societal coping range, and extreme events.**

Extreme heat, drought, and wildfires

As global temperatures increase, heatwaves increase. As precipitation becomes more variable and concentrated into more intense rainfall events, so dry periods get longer and droughts increase. The combination of extreme heat and drought causes more wildfires.

Heatwaves are often referred to as the 'silent killer'. They disproportionately affect the elderly, and it is sustained night-time temperatures that kill because while asleep older people are less able to regulate their body temperature. In the Lancet Countdown 2020 report, global heat-related mortality was tracked for people over 65 years of age since 1980. It showed a dramatic rise since 2010 in heat exposure of older people, driven by the combination of increasing heatwave occurrences and ageing populations. In 2019, there were a record 475 million exposure events, causing over 2.9 billion days of heatwave exposure for the elderly.

Heatwaves and droughts, however, are relative terms, as it depends where they occur and whether a region already has adaptations in place. The 2003 heatwave in Europe killed an estimated 70,000 people. Hardest hit was France with 14,800 deaths in the first three weeks of August and deaths in Paris increasing by 140%. After the 2003 heatwave it was realized that many of these deaths were due to the very weak public health response. As a result in many countries there were sweeping policy changes, including better heatwave prediction and emergency preparations, improved building design and air conditioning for hospitals and retirement homes, increased training for health professionals, an emphasis on responsible media coverage with health recommendations, and planned regular visits to the most vulnerable members of the population. These new policies throughout Europe have prevented a significant number of deaths during subsequent heatwaves, such as 2018. One of the reasons it is so difficult to understand the impacts of climate change is because people and societies can adapt to new conditions very quickly. Figure 25 shows the 2003 European heatwave in the context of summer temperatures over the last 100 years and those predicted for the next 100 years. What is clear is that the temperature of the 2003 heatwave could be the average summer temperature in 2050, and heatwaves above this new baseline may still occur. Adaptation to heatwaves, however, takes planning, resources, and money, and so though this has been possible in much of Europe there are many areas of the world where such preparation is not happening due to poverty and lack of good governance.

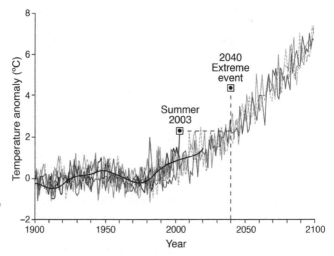

25. Comparison of the 2003 heat wave with past and future summer temperatures.

Droughts are also a major killer that should be considered. A drought happens when an area undergoes a prolonged period without sufficient water supply, whether surface or underground water. A drought can last for months or years and is usually caused when an area receives consistently below average rainfall. Droughts have a substantial impact on the local ecosystem and agriculture, including drops in crop growth and yield, and loss of livestock. Although droughts can persist for several years, even a short, intense drought can cause significant damage and harm to the local economy. Prolonged drought has caused famine, mass migrations, and humanitarian crises. From a disease point of view, droughts are much worse than floods because of a lack of fresh drinking water and stagnant pools of water bringing disease. In 2019, almost three times the global land surface area was affected by extended drought compared with the period between 1986 and 2005. One of the major concerns with climate change is that areas vulnerable to drought will have them more frequently, and areas that have never had them will start to experience them.

There has been an increase in wildfire risk in 114 out of 196 countries for the period 2016–19 as compared to the baseline period of 2001–4. Over this period there was a global increase to nearly 72,000 people per day being exposed to wildfire per year. Significant increases have occurred in Australia, Southern Hemisphere Africa, Brazil, and USA. The USA had record-breaking fire sessions in 2017, 2018, and 2020.

In Australia the 2019–20 bushfires season became known as the Black Summer. Throughout the summer, hundreds of fires burned, mainly in the south-east of the country, due to record heat and a prolonged drought. The fires burned an estimated 72,000 square miles, destroying nearly 10,000 buildings, and killing at least 450 people and 1 billion animals. Some endangered species may have been driven to extinction. With increasing incidences of extreme heat and droughts, wildfire risk will continue to grow across the world.

Storms and floods

Storms and floods are major natural hazards. Over the past two decades they have been responsible for three-quarters of the global insured losses, and over half the fatalities and economic losses from natural catastrophes. It is, therefore, essential that we know what is likely to happen in the future. There is evidence that the temperate regions, particularly in the Northern Hemisphere, have become more stormy over the past 50 years. Flood events have shown an upward trend since 2005, with almost three times the global land area flooded in 2019 (0.55–1.5% of global land surface area) compared with the period 1986–2005. This has not been driven by increases in any specific areas but by an increase in a wide range of events across the globe. The climate models suggest that the proportion of rainfall occurring as heavy rainfall has increased and will continue to do so, as will the year-to-year variability. This will increase the frequency and magnitude of flooding events.

Two-fifths of the world's population lives under the monsoon belt, which brings life-giving rains. Monsoons are driven by the temperature contrast between continents and oceans. Moisture-laden surface air blows from the Indian Ocean to the Asian continent, and from the Atlantic Ocean into west Africa during Northern Hemisphere summers, when the land masses become much warmer than the adjacent ocean. In winter, the continents become cooler than the adjacent oceans and high pressure develops at the surface, causing surface winds to blow towards the ocean. Climate models indicate an increase in the strength of the summer monsoons as a result of global warming over the next 100 years. There are three reasons why this should occur: (1) global warming will cause temperatures on continents to rise even higher than those of the ocean in summer and this is the primary driving force of the monsoon system; (2) decreased snow cover in Tibet, which is to be expected in a warmer world, will increase this temperature difference between land and sea, increasing the strength of the Asian summer monsoon; (3) a warmer climate means the air can hold more water vapour, so the monsoon winds will be able to carry more moisture. For the Asian summer monsoon, this could mean an increase of 10–20% in average rainfall, with an interannual variability of 25–100% and a dramatic increase in the number of days with heavy rain. The most concerning finding of the models is the predicted increase in rain variability between years, which could double, making it very difficult to predict how much rainfall will occur each year—essential knowledge for farmers.

One of the more contentious areas of climate change science is the study and predictions of future tropical cyclones, better known as typhoons or hurricanes. There is clear evidence that the number and intensity of hurricanes have increased over the past four decades in the North Atlantic Ocean, southern Indian Ocean, and Pacific Ocean. This is because the number and intensity of hurricanes are directly linked to the SST. As hurricanes can only start to form if the SST is above 26°C it would seem logical that in a warmer world there would be more hurricanes. Yet the actual formation of hurricanes is much rarer than the opportunities for them to occur. Only 10% of centres of falling pressure over the tropical oceans give rise to fully-fledged hurricanes. Other considerations, such as wind shear to start the rising air spinning, must be included to understand the genesis of tropical storms. In a year of high incidence, perhaps a maximum of fifty tropical storms will develop to hurricane levels. Predicting the level of a disaster is difficult, as the number of hurricanes is not the key, it is whether they make landfall, and how intense and prolonged they are once they have hit land.

When hurricanes hit in developed countries, the major effect is usually economic loss, while in developing countries it is loss of life. For example, Hurricane Katrina, which hit New Orleans in 2005, caused over 1,800 deaths and over $150 billion in damages. By contrast, Hurricane Mitch, which hit Central America in 1998, killed at least 11,000 people, made 1.5 million people homeless, and caused $6 billion in damages. And in 2013, Typhoon Haiyan, the most powerful tropical cyclone ever recorded, devastated large portions of South-East Asia, particularly the Philippines, affecting 11 million people, causing over 6,300 deaths with another 1,000 people missing, but only resulted in a reported $2.2 billion in damages.

One of the most important and mysterious elements in global climate is the periodic switching of the direction and intensity of ocean currents and winds in the Pacific. Originally known as El Niño ('the Christ child' in Spanish), as it usually appears at Christmas, and now more usually known as part of ENSO (El Niño–Southern Oscillation), this phenomenon typically occurs every three to seven years. It may last from several months to more than a year. ENSO is an oscillation between three climates: 'normal' conditions, La Niña (a cooler opposite of El Niño), and El Niño. El Niño conditions have been linked to changes in the monsoon, storm patterns, and occurrence of droughts all over the world. The 1997–8 El Niño conditions were the strongest on record and caused droughts in the southern USA, east Africa, northern India, north-east Brazil, and Australia. In Indonesia, forest fires burned out of control in the very dry conditions. In California, parts of South America, Sri Lanka, and east-central Africa, there were torrential rains and terrible floods.

The state of the ENSO has also been linked to the position and occurrence of hurricanes in the Atlantic. There is also considerable debate over whether ENSO has been affected by climate change. The El Niño conditions generally occur every two to six years, and in the past 40 years this has continued with no discernible pattern. Reconstruction of past climate using coral reefs in the western Pacific shows SST variations dating back 150 years, well beyond our historical records. The SST shows the shifts in ocean current which accompany shifts in the ENSO and reveals that there have been two major changes in the frequency and intensity of El Niño events. First was a shift at the beginning of the 20th century from a 10- to 15-year cycle to a 3- to 7-year cycle. The second was a sharp threshold in 1976 when a marked shift to more intense and frequent El Niño events occurred with a cycle of between two and four years. Climate models all agree that ENSO will continue in the future, and in the higher emission scenarios in the second half of this century, ENSO variability will become much more extreme, creating more rainfall and drought events, and influencing the number and intensity of tropical storms in unpredictable ways.

Coasts

As we have seen, the IPCC reports that sea level could rise by between 50 cm and 130 cm by 2100 compared with pre-industrial times. This prediction is of major concern to all those living in coastal areas, as rising sea levels will reduce the effectiveness of coastal defences against storms and floods, and increase the instability of cliffs and beaches. In the developed world, the response to this danger has been to add another few feet to the height of sea walls around property on the coast, the abandoning of some poorer quality agricultural land to the sea (as it is no longer worth the expense of protecting it), and the enhancement of legal protection for coastal wetlands, being nature's best defence against the sea. Globally, there are some nations based on small islands and river deltas that face a much more urgent situation (see Figure 26).

For small island nations, such as the Maldives in the Indian Ocean and the Marshall Islands in the Pacific, a 1 m rise in sea level would flood up to 75% of the dry land, making the islands uninhabitable. Interestingly, it is also these countries, which rely mainly on tourism, that have some of the highest fossil-fuel emissions per head of population. Other major concentrations of population at risk are those that live alongside river deltas, as in, for example, Bangladesh, Egypt, Nigeria, and Thailand. A World Bank report concluded that human activities on the deltas, such as dams and fresh-water extraction, were causing these areas to sink much faster than any predicted rise in sea level, increasing their vulnerability to storms and floods.

In the case of Bangladesh, over three-quarters of the country is within the deltaic region formed by the confluence of the Ganges, Brahmaputra, and Meghna rivers. Over half the country lies less than 5 m above sea level, so flooding is a common occurrence. During the summer monsoon a quarter of the country is flooded. Yet these floods, like those of the Nile, bring with them life as well

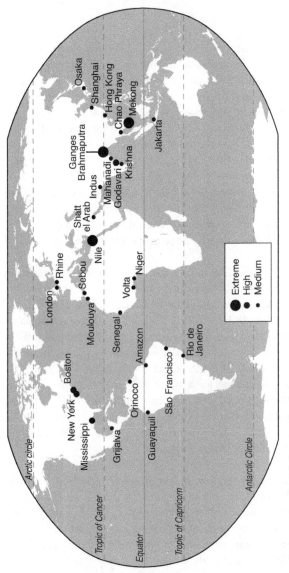

26. Areas most at risk from sea-level rise.

as destruction. The water irrigates and the silt fertilizes the land. The fertile Bengal delta supports one of the world's most dense populations, over 110 million people in 140,000 square kilometres (km^2). Every year, the Bengal delta should receive over 1 billion tonnes of sediment and 1,000 cubic kilometres (km^3) of fresh water. This sediment load balances the erosion of the delta by both natural processes and human activity. However, the Ganges, Brahmaputra, and Meghna have been dammed for irrigation and power generation, preventing the movement of silt downriver. The reduced sediment input is causing the delta to subside. Exacerbating this is the rapid extraction of fresh water.

Since the 1980s, 100,000 tube wells and 20,000 deep wells have been sunk, increasing the fresh-water extraction six-fold. These wells are essential to improving the quality of life for people in this region, but they have produced a subsidence rate of up to 2.5 cm per year, one of the highest rates in the world. Using estimates of subsidence rate and global warming sea-level rise, the World Bank has estimated that by the end of the 21st century, the relative sea level in Bangladesh could rise by as much as 1.8 m. In a worst-case scenario, they estimated that this would result in a loss of up to 16% of land, supporting 13% of the population, and producing 12% of the current GDP. Unfortunately, this scenario does not take any account of the devastation of the mangrove forest and the associated fisheries. Moreover, increased landward intrusions of salt water would further damage water quality and agriculture.

Many major cities around the world are vulnerable to flooding because they were built close to rivers or the coast in order to facilitate trade via the oceans. Examples of current cities most at risk include, in Asia: Dhaka (20.3 million people today), Shanghai (17.5 million), Guangzhou (13 million), Shenzen (12.5 million), Jakarta (10.8 million), Bangkok (10.5 million), Hong Kong (8.4 million), Ho Chi Minh City (8.3 million), and Osaka (5.2 million); in North America: New York (18.8 million), Boston (4.9 million), Miami (2.7 million), and New Orleans (0.4 million);

in South America: Guayaquil (2.9 million) and Rio de Janeiro (1.8 million); in Africa: Abidjan (3.7 million) and Alexandria (3.0 million); and in Europe: London (8.9 million) and The Hague (2.5 million).

Consider the case of London. At the moment, London is protected from flooding by the Thames Barrier. The Thames Barrier was built in response to the catastrophic floods of 1953, and was finally ready for use in 1982 (it was officially opened on 8 May 1984). The Thames Barrier protects 150 km^2 of London and property worth at least £1.5 trillion. Because of the foresight of previous scientific advisors to the UK government, it was built to withstand a 1-in-2,000-year flood. With the increased sea level due to climate change, this protection by 2030 will drop to a 1-in-1,000-year event. By 2020 the Barrier had closed 193 times in its 38-year history; and over 40% of these closures have occurred in the last 10 years. The UK economy is the sixth largest in the world, with approximately £1.4 trillion per year generated through London. London is also one of the three main centres, along with New York and Tokyo, for 24-hour share-trading. If London were disabled by a major flood then not only would this hit the economy of the UK, but potentially it could disrupt global trade. Hence the UK Environment Agency has put in place plans to guard against a significant sea-level rise in the future, including plans for a new barrier between Essex and Kent to guard against a possible 4 m rise in sea level. But most other cities around the world do not have the resources to plan for this sort of protection.

The Lancet Countdown 2020 report estimates that without intervention, between 145 million and 565 million people could be affected and displaced as a result of future sea-level rise.

Agriculture

One of the major worries concerning climate change is the effect it will have on agriculture, both globally and regionally. The main question is whether the world can feed itself with an extra 2 billion people on the planet by 2050 and a rapidly changing climate. Figure 27 shows the drop in cereal grain yields that has already occurred. Modelling suggests that in higher latitudes agricultural productivity may increase due to the longer growing season and reduced frost damage—but some of this will be offset by more frequent crop damage due to extreme weather events. Agriculture production will however reduce significantly in the tropics and sub-tropics due to much hotter temperatures and more variable rainfall.

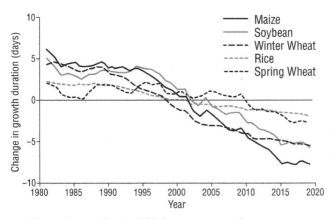

27. **Changes in cereal grain yields between 1980 and 2020.**

Higher temperatures and humidity will also be a challenge for many societies that rely heavily on subsistence agriculture, as higher air temperatures and humidity make working outside more difficult and increase the likelihood of hyperthermia. This will impact on the health of anyone who has to work outside regularly, including construction and farm workers. Across the globe already a potential 278 billion work hours were lost in 2019 due to extreme conditions—92 billion hours greater than in 2000. Seven countries (Cambodia, India, China, Indonesia, Nigeria, Brazil, and the USA) together represent approximately 60% of the global hours lost in 2019, with India experiencing by far the greatest loss. In the first six countries, the impact on work hours lost fell mainly on agricultural workers.

Estimating the overall impact of climate change on agriculture is difficult as agricultural production has very little to do with feeding the world's population and much more to do with trade and economics. This is why the European Union has stockpiles of food, while many underdeveloped countries export cash crops (such as sugar, cocoa, coffee, tea, and rubber) but cannot adequately feed their own populations. A classic example is the west African state of Benin, where cotton farmers can obtain cotton yields four to eight times per hectare greater than their US competitors in Texas. However, because the USA subsidizes its farmers, this means that US cotton is cheaper than that coming from Benin. Currently, US cotton farmers receive over $4 billion in subsidies, almost twice the total GDP of Benin. In 2002, Brazil filed a case with the World Trade Organization (WTO) against the USA for unfair subsidies and distortion of trade. They won their case in 2005; however, 15 years later, the USA is still discussing what changes should be made to their farming subsidies. So even if climate change makes Texan cotton yields even lower, it does not change the biased market forces still illegally in operation.

Markets can reinforce the difference between agricultural impacts in developed and developing countries. Variations in supply and demand could mean that agricultural exporters may gain in monetary terms even if the supplies fall, because when a product becomes scarce, the price rises. The other completely unknown factor is the extent to which a country's agriculture can be adapted. For example, in climate change models it is assumed that production levels in developing countries will fall to a greater degree than those in developed countries because their estimated capability to adapt is less than that of developed countries. But this is just another assumption that has no analogue in the past, and as these effects on agriculture will occur over the next century, many developing countries may catch up with the developed world in terms of adaptability.

One example of the real regional problems that climate change could cause is the case of coffee growing in Uganda. Here, the total area suitable for growing Robusta coffee would be dramatically reduced, to 10% of the present area, by a temperature increase of 2°C. Only higher areas of land would remain suitable for coffee growing, the rest would become too hot. But no one can tell whether these remaining areas would make more or less money for the country, because if other coffee growing areas around the world are similarly affected, the price of coffee beans will increase due to scarcity. This demonstrates the vulnerability to the effects of global warming of many developing countries whose economies often rely heavily on one or two agricultural products, as it is very difficult to predict the changes that global warming will cause in terms of crop yield and its cash equivalent. Hence one major adaptation to global warming should be the broadening of the economic and agricultural base of the most threatened countries. This, of course, is much harder to accomplish in practice than on paper, and it is clear that the USA, EU, and China agricultural subsidies and the current one-sided world trade agreements have a greater effect on global agricultural production and the ability of countries to feed themselves than climate change will ever have. Solutions look to be even further away with the failure of the WTO negotiations.

Ocean acidification

Direct measurements of the ocean's chemistry have shown that the pH of the oceans is getting lower; that is, they are getting more acidic (see Figure 28). This is because CO_2 in the atmosphere dissolves in the surface water of the ocean. This process is controlled by two main factors: the amount of CO_2 in the atmosphere and the temperature of the ocean. The oceans have already absorbed about a third of the CO_2 resulting from human activities, which has led to a steady decrease in ocean pH levels. With increasing atmospheric CO_2 in the future, the amount of dissolved CO_2 in the ocean will continue to increase. Some marine organisms, such as corals, foraminifera, coccoliths, and shellfish, have shells composed of calcium carbonate, which dissolves in acid. Laboratory and field experiments show that under high CO_2 levels the more acidic waters cause some marine species to develop misshapen shells and have lower growth rates, although the effect varies among species. Acidification also alters the cycling of nutrients and many other elements and compounds in the ocean, and it is likely to shift the competitive advantage among species, and have impacts on marine ecosystems and the food web. This is a major worry as fishing is still an important source of food, with about 95 million tonnes of fish caught by commercial fishing and another 50 million tonnes produced by fish farms per year.

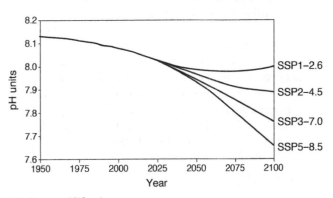

28. Ocean acidification.

Biodiversity

The current loss of biodiversity around the world is due to human activity, including deforestation, agriculture, urbanization, and mineral exploitation. Extinction rates are currently 100–1,000 times higher than the background natural rate and climate change will exacerbate this decline. The IPCC impact report lists the following species as those most at risk from climate change: the mountain gorilla in Africa; amphibians that live in the cloud forests of the neotropics; the spectacled bear of the Andes; forest birds of Tanzania; the 'resplendent quetzal' bird in Central America; the Bengal tiger, and other species found only in the Sundarban wetlands; rainfall-sensitive plants found only in the Cape Floral Kingdom of South Africa; and polar bears and penguins near the poles. The primary reason for the threat to these species is that they are unable to migrate in response to climate change because of their particular geographical location or the encroachment of human activity, especially farming and urbanization. An example of the former is the cloud forests of the neotropics: as climate changes, this particular climatic zone will migrate up the mountainside—to the point where there is no more mountain to climb.

One example of an ecosystem under threat is the coral reefs. Coral reefs are a valuable economic resource for fisheries, recreation, tourism, and coastal protection. Some estimate that the global cost of losing the coral reefs runs into hundreds of billions of dollars each year. In addition, reefs are one of the largest global stores of marine biodiversity. The last few years have seen unprecedented declines in the health of coral reefs. An estimated 50% of the hard corals on Australia's Great Barrier Reef have been lost in the past few years from bleaching events, which are caused by extreme water temperatures. The Great Barrier Reef is the world's largest coral reef system, composed of over 2,900 individual reefs and 900 islands stretching for over 1,400 miles. In other regions, as much as 70% of the coral has died in a single season. There has also been an upsurge in the variety, incidence, and virulence of coral disease in recent years, with major die-offs in Florida and much of the Caribbean region. In addition, increasing atmospheric CO_2 concentrations could decrease the calcification rates of the reef-building corals, resulting in weaker skeletons, reduced growth rates, and increased vulnerability to erosion. Model results suggest these effects would be most severe at the current margins of coral reef distribution.

On a more theoretical note, the biologist Chris Thomas and colleagues published a study in *Nature* investigating the possible increase in extinction rates over the next 50 years in key regions such as Mexico, Amazonia, and Australia. The theoretical models suggest that a 2 °C warming by 2050 could condemn to extinction one-quarter of all species they studied. This study has been criticized as there are many assumptions in their models which may or may not be true: for example, they assume we know the full climatic range in which each species can persist and the precise relationship between shrinking habitat and extinction rates. So these results should be seen only as the likely direction of extinction rates, not necessarily the exact magnitude. However, this and many other scientific studies demonstrate the huge threat to regional and global biodiversity and illustrate the sensitivity of biological systems to the amount and rate of warming that will occur in the future.

Protecting biodiversity has other major benefits for human society. In 2020, the world was brought to a standstill by the Covid-19 pandemic. One of the reasons that Covid-19 is such a complex, severe, and even fatal respiratory disease is that it is a zoonotic virus, a virus that has jumped from another animal and mutated to infect humans. Hence the virus has a genetic signature unknown to our immune systems, delaying our ability to develop antibodies that can fight the infection. It seems increasingly likely that it is the illegal trade in endangered animals such as bats and pangolins through inhumane 'wet markets' in China and South-East Asia that allowed its transmission between species. The extremely high risks of such zoonotic virus outbreaks were indicated by previous outbreaks, such as avian influenza related to the H5N1 virus in 1996 and the SARS outbreak in 2002–3. Both times, Chinese wet markets were temporally banned and then allowed to continue. Hence there is a real need to protect and respect biodiversity and wildlife to prevent these zoonotic diseases occurring again. The Chinese and other governments need to promote cultural change, along with gradual regulatory restrictions to protect wildlife and consequently humans too.

Human health

The potential health impacts of climate change are immense, and managing those impacts will be an enormous challenge. Climate change will increase deaths from heatwaves, droughts, wildfires, storms, and floods. Higher temperatures and variable precipitation threaten food production. This is due to reduced productivity because of the increased risks to people working outside regularly, such as construction and farm workers. It has been suggested that the overall death rate may drop in some countries since many elderly people die from cold weather so warmer winters would reduce this cause of death. However, this view has shown to be incorrect, as recent research has demonstrated that better housing, improved health care, higher incomes, and greater awareness of the risks of cold have been responsible for the reduction in winter deaths in the UK since 1950, while in the USA summer-heat-related deaths have been four times higher than deaths from cold for the past three decades. Hence in many societies adaptation to cold climates and improved protection for the most vulnerable members of society means that warmer winters will have little or no effect in reducing the death rate.

The 2009 University College London's report in the *Lancet* journal, 'Managing the Health Effects of Climate Change', identified the two major areas that could affect the health of billions of people: water and food. The most important threat to human health is lack of access to fresh drinking water. At present there are still 1 billion people who do not have regular access to clean, safe drinking water. Not only does the lack of water cause major health problems from dehydration, but a large number of diseases and parasites are present in dirty water. The rising worldwide human population, particularly those concentrated in urban areas, is putting a great strain on water resources. The impacts of climate change—including changes in temperature, precipitation, and sea levels—are expected to have varying consequences for the availability of fresh water around the world. For example, changes in river run-off will affect the yields of rivers and reservoirs, and thus the recharging of groundwater supplies. An increase in the rate of evaporation will also affect water supplies and contribute to the salination of irrigated agricultural lands. Rising sea levels may result in saline intrusion in coastal aquifers. Currently, approximately 2 billion people, one-quarter of the world's population, live in countries that are water-stressed. It has been suggested that if nothing is done to mitigate climate change then up to 50% of the world population could live in countries experiencing water-stress by 2050. Of these people, 80% will be in developing countries.

Climate change is likely to have the greatest impact in countries with a high ratio of relative use to available supply. Regions with abundant water supplies will get more than they want with increased flooding. As suggested above, computer models predict much heavier rains and thus major flood problems for Europe, whilst, paradoxically, countries that currently have little water (e.g. those relying on desalination) may be relatively unaffected. It will be countries in between, with no history or infrastructure for dealing with water shortages, that will be the most affected. In central Asia, and northern and southern Africa there will be even less rainfall, and water quality will become increasingly degraded through higher temperatures and pollutant run-off. Add to this the predicted increase in year-to-year variability in rainfall, and droughts will become more common. Hence it is those countries that have been identified as most at risk which need to start planning now to conserve their water supplies and/or to deal with the increased risks of flooding; because it is the lack of infrastructure to deal with drought and floods rather than the lack or abundance of water which causes the threat to human health.

Food security rests on three main pillars: (1) food availability—is enough being produced? (2) Access—can people afford it? And (3) stability—is there always food available? According to the UN World Food Programme, we currently produce enough food to feed 10 billion people, easily enough to cover the predicted increase in population this century. But there are 821 million people on the brink of starvation today, up by 25 million in just five years. This is because they simply do not have enough money to buy food. Climate change threatens food availability and stability as it affects the production of food and other agricultural goods. Extreme weather events must also be considered. With an increasingly globalized economy very few countries are self-sufficient in basic food and hence food imports are very important. The cost of basic food items can be strongly influenced by global demand, national agricultural subsidies and export bans, and natural disasters, but the biggest influence is food speculation on the global markets. In 2008–9, there was a 60% rise in the price of food and in 2011–12 there was a 40% jump in price, both linked to food speculation. So the inability of many people to afford basic food, leading to malnutrition and starvation, can be linked directly to speculation on food prices on the global markets in London, New York, and Tokyo.

Another threat to human health is the transmission of infectious diseases, which is directly affected by climatic factors. Climate change will particularly influence vector-borne diseases, that is, diseases that are carried by another organism—for example, malaria, which is carried by mosquitoes. Infective agents and their vector organisms are sensitive to factors such as temperature, surface-water temperature, humidity, wind, soil moisture, and changes in forest distribution. It is, therefore, projected that climate change and altered weather patterns would affect the range (both altitude and latitude), intensity, and seasonality of many vector-borne and other infectious diseases. For example, there is a strong correlation between increased SST and sea level, and the severity of the cholera epidemics in Bangladesh. With predicted future climate change and consequent rise in Bangladesh's relative sea level, cholera epidemics could become more common.

In general, then, increased warmth and moisture caused by climate change will enhance transmission of diseases. But while the potential transmission of many of these diseases increases in response to climate change, we should remember that our capacity to control the diseases will also change. New or improved vaccination can be expected; some vector species can be constrained by the use of pesticides. Nevertheless, there are uncertainties and risks here, too: for example, long-term pesticide use encourages the breeding of resistant strains, while killing many natural predators of pests.

The most important vector-borne disease is malaria, with currently 500 million infected people worldwide. *Plasmodium vivax*, which is carried by the *Anopheles* mosquito, is the organism that causes malaria. The main climate factors that have a bearing on the malarial transmission potential of the mosquito population are temperature and precipitation. Assessments of the potential impact of global climate change on the incidence of malaria suggest a widespread increase of risk because of the expansion of the areas suitable for malaria transmission. Already in the past five years, the area suitable for malaria transmission in highland areas was 39% higher in Africa and 150% higher in east Asia compared to the 1950s. Mathematical models mapping out the temperature zones suitable for mosquitoes suggest that by the 2080s the potential exposure of people could increase by 2–4% (260–320 million people). The predicted increase is most pronounced at the borders of endemic malarial areas and at higher altitudes within malarial areas. The changes in malaria risk must be interpreted on the basis of local environmental conditions, the effects of socioeconomic development, and malaria control programmes or capabilities. Climate change will also provide excellent conditions for *Anopheles* mosquitoes to breed in southern England, continental Europe, and the northern USA.

It should, however, be noted that the occurrence of many tropical diseases is related to development. As recently as the 1940s, malaria was endemic in Finland, Poland, Russia, and thirty-six states in the USA including Washington, Oregon, Idaho, Montana, North Dakota, New York, Pennsylvania, and New Jersey. So although climate change has the potential to increase the range of many of these tropical diseases, the experience of Europe and the USA suggests that combating malaria is strongly linked to development and resources: development to ensure efficient monitoring of the disease and resources to secure a strong effort to eradicate the mosquitoes and their breeding grounds.

The impacts of climate change will increase significantly as the temperature of the planet rises. Climate change will affect the return period and severity of heatwaves, droughts, wildfires, storms, and floods. Coastal cities and towns will be especially vulnerable as sea levels rise, increasing the impacts of floods and storm surges. Water and food security as well as public health will become the most important problems facing all countries. Climate change threatens global biodiversity and the wellbeing of billions of people. In Table 4 I have tried to summarize the potential impacts of climate change. Though many colleagues are planning on how to deal with a 4°C world, my simple advice is, let us not go there.

Table 4. Potential impacts of climate change

Temperature rise above pre-industrial levels	Potential impacts of climate change
1.5°	• Major effects on warm-water coral reef ecosystem. • Significant impacts on vulnerable ecosystems and species (polar regions, wetlands, and cloud forests). • Increase in coastal and river flooding. • Increase in extreme weather events. • Increase in the spread of tropical infectious disease. • Increase in heat-related morbidity and mortality.
2°–3°C	• The Maldives, the Marshall Islands, Tuvalu, and many other small island nations abandoned. • Major loss of warm-water coral reef ecosystem. • Major changes in the Arctic regions with a substantial loss of Arctic sea ice. • Major increase in extreme weather events and the spread of infectious disease. • Major increase in heat-related morbidity and mortality, especially in the low latitudes. • Significant impacts on vulnerable ecosystems (polar regions, wetlands, cloud forests, and mangroves). • Significant increase in coastal and river flooding around the world. • Significant impacts on low latitude fisheries. • Decrease in crop yields and productivity especially in the tropics and sub-tropical regions.

3°–4°C	• Major impacts on all ecosystems including significant increase in species extinctions.
	• Loss of all warm-water and many cold-water coral reef ecosystems.
	• Arctic completely free of sea ice in summer, Arctic temperature increase by 8°C.
	• Disappearance of the majority of mountain glaciers, including all ice on Kilimanjaro (Tanzania).
	• Major increase in extreme weather events and spread of infectious disease.
	• Major decreases in agricultural and fishery production and available water resources.
	• Food and water security become major political and humanitarian issues.
	• Environmental forced mass migration increases.
	• Ocean and terrestrial carbon sinks reduce, accelerating climate change.
4°–5°C	• Catastrophic loss of ecosystems and species all around the world.
	• Melting of western Antarctic and Greenland ice sheets accelerates, causing significant rises in global sea level.
	• Fifth of world population affected by flooding and major coastal cities abandoned.
	• Environmental forced mass migration accelerates and conflicts over resources increase.
	• In many countries, summer temperatures persistently stay above 40°C.
	• Heatwaves with temperatures as high as 50°C become common.
	• Over 3.5 billion people become water-stressed.
	• Wildfires create major air pollution events and human health crises.
	• Global food production plummets, leading to widespread malnutrition and starvation.
5°–6°C or higher	• Don't go there.

Chapter 6
Climate surprises

We are changing the composition of the atmosphere beyond anything that has been experienced over the past 3 million years. We are headed for unknown territory and therefore scientific uncertainty can be great. We know from the study of past records that the climate system can switch into a new state very quickly once a threshold has been passed. For example, ice-core records suggest that half the warming in Greenland at the end of the last ice age was achieved in only a few decades. This chapter examines the possibility that there are thresholds or tipping points in the climate system that may occur as we warm the planet. Figure 29 shows the main tipping points which scientists have been concerned about over the last three decades. Irreversible melting of the Greenland and/or western Antarctic ice sheet, slowing down of the North Atlantic deep-ocean circulation, massive release of CH_4 from melting gas hydrates, and the Amazon rainforest dieback will all be discussed.

Thresholds and tipping points

The relationship between a climate forcing factor such as GHGs and the climate response is complicated. In an ideal world it would be a simple relationship with little or no delay, but we already know that there is inertia in the climate system, so that it responds to GHG forcing with a 10- to 20-year delay, depending

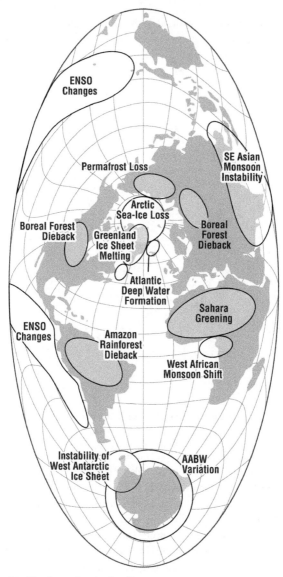

29. Tipping points in the climate system.

on how much is being emitted. So we can examine the way
different parts of the climate system respond to climate change
with four scenarios (see Figure 30):

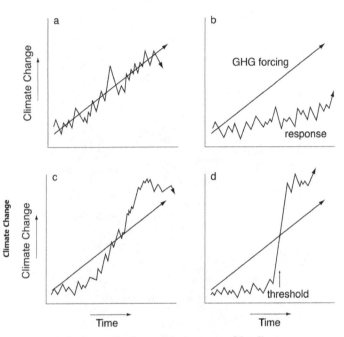

30. Greenhouse gas forcing and the response of the climate system.

(a) *Linear but delayed response* (Figure 30a). In this case, the
increase in GHGs produces a delayed but direct response in the
climate system whose magnitude is in proportion to the
additional forcing. This can be equated to pushing a car along a
flat road. At first nothing happens as friction must be overcome
before the car will start moving. Once that has happened most of
the energy put into pushing is used to move the car forward. An
example of this is the heating up of the oceans which has a delay
due to the inertia of heating such a large body of water.

(b) *Muted or limited response* (Figure 30b). In this case, the GHG forcing may be strong, but the relevant part of the climate system is in some way buffered and therefore gives very little response. This is analogous to pushing the car up a hill: you can spend as much energy as you like trying to push the car, but it will not move very far. An example of this is the east Antarctic ice sheet, which has been stable at much warmer temperatures than today.

(c) *Delayed and non-linear response* (Figure 30c). In this case, the climate system may have an initial slow response to the GHG forcing but then respond in a non-linear way. This is a real possibility when it comes to climate change if we have underestimated the positive feedback in the system. This scenario can be equated to the car near the top of a hill: it takes some effort and thus time to push the car to the very top of the hill; this is the buffering effect. Once the car has reached the peak, it takes very little effort to push the car over it, and then it accelerates down the hill with or without help. Once it reaches the bottom, the car then continues for some time—the overshoot—and then it slows down of its own accord and settles into a new state.

(d) *Threshold response* (Figure 30d). In this case, initially, there is very little response in that part of the climate system to the GHG forcing. However, on reaching a threshold, all the response takes place in a very short period of time, in a single, large step. In many cases, the response may be much greater than one would expect from the size of the forcing, and this can be referred to as a 'response overshoot'. This scenario equates to the bus hanging off the cliff at the end of the original film *The Italian Job*: as long as there are only very small changes, nothing happens at all. However, a critical point (in this case weight) is reached and the bus (and the gold) plunges off the cliff into the ravine below. An example of this could be the Greenland ice sheet that has started to melt; the melting could suddenly accelerate, causing a catastrophic collapse.

There are also situations in which a threshold becomes a tipping point. You can think of a threshold as a point at which there is change in a system that can be reversed. But a tipping point is a threshold that, when crossed, means the system moves into a new state and this transition is irreversible. An added complication when assessing whether climate change will create a simple threshold or a tipping point are bifurcations within the climate system. This means the forcing required to push the climate system one way across the threshold is different from the reverse. This implies that once a climate threshold has been passed, it is a lot more difficult to reverse it and in some cases it may in effect be irreversible.

The term 'tipping points' is used a lot in climate change research and discussions. However, care must be taken as there are two usages of this word. First, there are references to climate tipping points, which are the large-scale, irreversible shifts in the climate system, such as irreversible melting of ice sheets or the release of huge stores of CH_4 from below the oceans. The other usage concerns societal tipping points, which occur when climate change has a major effect on a region or a particular country. For example, a 200 mile (~322 km) shift northward of the South-East Asian monsoonal rainfall belt is, in climatological terms, a small shift, not a fundamental climate tipping point. But for the countries where the rains no longer fall or those where it does for the first time, such a shift is a major climate tipping point, because their weather may have been permanently altered.

Melting ice sheets

The IPCC projections for sea-level rise by 2100, if there are no significant curbs to carbon emissions, are between 0.50 m and 1.3 m. The largest uncertainty within these estimates is the contribution that the melting of Greenland and Antarctica will make by the end of the century. At the moment it is estimated that Greenland is losing over 230 Gt of ice per year, a seven-fold increase since the early 1990s. Antarctica is losing about 150 Gt of ice per year, a five-fold increase since the early 1990s. Most of this loss is from the northern Antarctic peninsula and the Amundsen sea sector of west Antarctica. Greenland and Antarctica together constitute one of the most worrying potential climate surprises. If the large ice sheets there melted completely, their contribution to global sea-level rise would be as follows: Greenland, about 7 m; the west Antarctic ice sheet, about 8.5 m; and the east Antarctic ice sheet, about 65 m. These compare with just 0.3 m if all the mountain glaciers melted. Palaeoclimate data show that the huge east Antarctic ice sheet developed 35 million years ago due to the progressive tectonic isolation of Antarctica and that it has in fact remained stable in much warmer climates. So climate scientists have a very high degree of confidence that the east Antarctic ice sheet will remain stable in this century.

However, scientists are very worried that the melting of Greenland or west Antarctica could significantly accelerate in the next 100 years. Even if we have already started the processes of melting the whole of these ice sheets, there is a physical constraint to the speed at which the ice can melt. This is due to the time it takes for heat to penetrate the ice sheets. Imagine dropping an ice cube into a hot cup of coffee. You know it will melt entirely, but it takes time for the heat to penetrate to the middle of the ice cube. Most of the ice from ice sheets flows through ice streams to get to the sea, and there is a limit on how much ice these ice streams can transport. The worst-case scenario, according to leading glaciologists, is that these ice sheets could add between 1 m and 1.5 m to the sea level by the end of the century, which would threaten many coastal populations around the world. There is also scientific debate about what happens to both the Greenland and Antarctic ice sheets beyond the next 100 years. Even if significant melting does not occur this century, we may have started a process that causes irreversible melting during the next one. Our carbon emissions over the next few decades could determine the long-term future of the ice sheets and the livelihoods of billions of people who live close the coast.

Deep-ocean circulation

The circulation of the ocean is one of the major controls on our global climate. In fact, the deep ocean is the only candidate for driving and sustaining internal long-term climate change (of hundreds to thousands of years) because of its volume, heat capacity, and inertia. In the North Atlantic, the north-east trending Gulf Stream carries warm and salty surface water from the Gulf of Mexico up to the Nordic seas. The increased saltiness, or salinity, in the Gulf Stream is due to the huge amount of evaporation that occurs in the Caribbean, which removes moisture from the surface waters and concentrates the salts in the sea water. As the Gulf Stream flows northward, it cools down. The combination of a high salt content and low temperature increases the density of the surface water. Hence, when it reaches the relatively low saline oceans north of Iceland, the surface water has cooled sufficiently to become dense enough to sink into the deep ocean. The 'pull' exerted by the sinking of this dense water mass helps maintain the strength of the warm Gulf Stream, ensuring a current of warm tropical water continues to flow into the north-east Atlantic, sending mild air masses across to the European continent. It has been calculated that the Gulf Stream delivers 27,000 times the energy of all of Britain's power stations put together. If you are in any doubt about how good the Gulf Stream is for the European climate, compare the winters at the same latitude on either side of the Atlantic Ocean, for example London with Labrador, or Lisbon with New York. Or a better comparison is between Western Europe and the west coast of North America, which have a similar geographical relationship between the ocean and continent—so think of Alaska and Scotland, which are at about the same latitude.

The newly formed deep water sinks to a depth of between 2,000 m and 3,500 m in the ocean and flows southward down the Atlantic Ocean, as the North Atlantic Deep Water (NADW). In the South Atlantic Ocean, it meets a second type of deep water, which is formed in the Southern Ocean and is called the Antarctic Bottom Water (AABW). This is formed in a different way to NADW. Antarctica is surrounded by sea ice and deep water forms in coast polynyas, or large holes in the sea ice. Out-blowing Antarctic winds push sea ice away from the continental edge to produce these holes. The winds are so cold that they super-cool the exposed surface waters. This leads to more sea-ice formation and salt rejection, producing the coldest and saltiest water in the world. AABW flows around the Antarctic and penetrates the North Atlantic, flowing under the warmer, and thus somewhat lighter, NADW (see Figure 31a). The AABW also flows into both the Indian and Pacific Oceans.

This balance between the NADW and AABW is extremely important in maintaining our present climate, as not only does it keep the Gulf Stream flowing past Europe, but it also maintains the right amount of heat exchange between the Northern and Southern Hemispheres. Scientists have shown that the circulation of deep water can be weakened or 'switched off' if there is sufficient input of fresh water to make the surface water too light to sink. This evidence has come from both computer models and the study of past climates. Scientists have coined the phrase 'dedensification' to mean the removal of density by the addition of fresh water and/or warming of the water, both of which prevent sea water from being dense enough to sink. As we have seen, there is already concern that global warming will cause significant melting of the polar ice caps. This will lead to more fresh water being added to the polar oceans. Climate change could, therefore, cause the collapse of the NADW, and a weakening of the warm Gulf Stream (Figure 31b). This would cause much colder European winters and more severe weather. However, the influence of the warm Gulf Stream is mainly seen in the winter and has only a small effect on summer temperatures. So, if the Gulf Stream fails, global warming would still cause European summers to heat up. Europe would end up with extreme seasonal weather, very similar to that of Alaska.

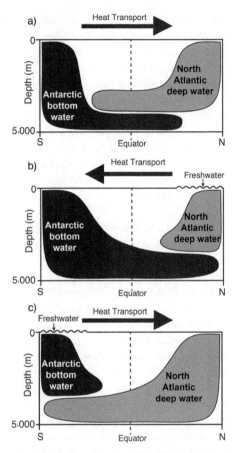

31. Deep-ocean circulation changes depending on freshwater inputs.

A counter-scenario is that, if the Antarctic ice sheet starts to melt significantly before the Greenland and Arctic ice, things could be very different. If enough melt-water goes into the Southern Ocean, then AABW will be severely curtailed. Since the deep-water system is a balancing act between NADW and AABW, if AABW is reduced then the NADW will increase and expand (Figure 31c). The problem is that NADW is warmer than AABW and, because if you heat up a liquid it expands, the NADW will take up more space. So any increase in NADW could mean a rise in sea level. Computer models by Dan Seidov (now at the National Oceanic and Atmospheric Administration) and myself have suggested that such a scenario would result in an average sea-level increase of over 1 m.

It has been over 30 years since the possibility of a catastrophic shut down of the deep-ocean circulation was suggested, and there has been a huge amount of work on it. Monitoring has shown that the Gulf Stream has weakened by 15% since the middle of the last century. Evidence collated from this ocean monitoring and climate model predictions of the future in the very latest IPCC report suggest that collapse of the Gulf Stream is highly unlikely in the 21st century. The models do, however, show a significant weakening in the overturning of the North Atlantic in this century, especially in the high-emission scenarios, and the problem is that we do not know where a potential tipping point leading to shut down of deep-ocean circulation might be. Moreover, if the melting of Greenland or western Antarctica accelerates, then huge amounts of fresh water could enter the oceans, significantly disrupting deep-ocean circulation.

Gas hydrates

Below the world's oceans and permafrost is stored a large amount of carbon in the form of CH_4. The CH_4 gas is trapped in a solid cage of water molecules at low temperatures and/or high pressures. The CH_4 gas comes from decaying organic matter found deep in ocean sediments and in soils beneath permafrost (Figure 32). These gas hydrate reservoirs could be unstable, as an increase in temperature or decrease in pressure would cause them to destabilize and to release the trapped CH_4. Climate change is warming up both the oceans and the permafrost, threatening the stability of gas hydrates. CH_4 is a strong GHG, twenty-one times

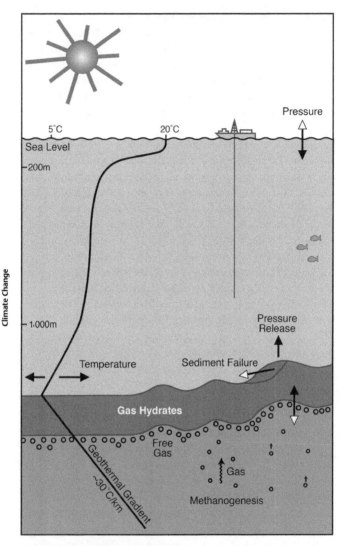

32. Gas hydrates in a marine setting.

more powerful than CO_2 (see Table 1). If enough were released, it would raise global temperatures, which could lead to the release of even more gas hydrates—producing a runaway effect. Scientists really have no idea how much CH_4 is stored in the gas hydrates beneath our feet: estimates are of between 1,000 and 10,000 Gt (compared with ~800 GtC currently in the atmosphere), which is a huge range. Without a more precise estimate, it is very difficult to assess the risk posed by gas hydrates.

The reason why scientists are so worried about this issue is because there is evidence that a super-greenhouse effect occurred 55 million years ago, during what is called the Palaeocene–Eocene Thermal Maximum (PETM). During this hot-house event, scientists think that up to 1,500 Gt of gas hydrates may have been released. This huge injection of CH_4 into the atmosphere accelerated the natural greenhouse effect, producing an extra 5°C of warming. There is still, however, considerable debate over the PETM. For example, was it gas hydrate CH_4 release or CO_2 release from a phase of massive volcanism occurring around the same time that was the main cause of the warming?

The current consensus is that the ocean reserves of gas hydrate are likely to remain stable this century. Gas hydrates form a solid layer at the bottom of the ocean. The depth of this layer is controlled by the geothermal heat gradient—as you go deeper in the sediment, the temperature rises at about 30°C/km. At a certain depth, it is too warm for gas hydrates to exist and CH_4 collects there as free gas in the sediment. As ocean temperatures change, the temperature change has to be transmitted through the solid gas hydrate layer to the lower boundary for some of it to melt. If this process is slow enough, the gas released migrates upwards in the ocean sediment column and refreezes at a higher level. However, if carbon emissions are not curbed, then by the next century we could see this process speed up, leading to the release of some of the CH_4 stored in the deep ocean.

It is clear that the gas hydrate below what was once permafrost is already melting, with bubbles observed in many Canadian and Siberian lakes. With the Arctic amplification temperature, rises will be nearly twice the global average in the northern polar regions, which will accelerate the gas hydrate melting. But we still do not have an indication of how much CH_4 is stored beneath the world's permafrost regions. So at the moment, our best estimate suggests a global warming of 3°C could release between 35 and 940 GtC, which could add between 0.02°C to 0.5°C to global temperatures.

Amazon dieback

In 1542, Francisco de Orellana led the first European voyage down the Amazon River. During this intrepid journey the expedition met with a lot of resistance from the local Indians; in one particular tribe the women warriors were so fierce that they drove their male warriors in front of them with spears. Thus the river was named after the famous women warriors of the Greek myths, the Amazons. This makes Francisco dexOrellana one of the unluckiest explorers of that age, as ordinarily the river would have been named after him. The Amazon River discharges approximately 20% of all fresh water carried to the oceans. The Amazon drainage basin is the world's largest, covering an area of 7,050,000 km²—about the size of Europe. The river is a product of the Amazon monsoon, which every summer brings huge rains. This also produces the spectacular expanse of rainforest, which supports the greatest diversity and largest number of species of any area in the world.

The Amazon rainforest is important when it comes to climate change as it is a huge natural store of carbon. Originally it was thought that established rainforests such as the Amazon had reached maturity. Detailed surveys of all the rainforests of the world over the past four decades show this is incorrect. In the 1990s, intact tropical forests—those unaffected by logging or fires—removed roughly 46 billion tonnes of CO_2 from the atmosphere. The sting in the tail is that this removal had diminished to an estimated 25 billion tonnes in the 2010s. The lost sink capacity is 21 billion tonnes of CO_2, equivalent to a decade of fossil-fuel emissions from the UK, Germany, France, and Canada combined. All this data is compiled by the African Tropical Rainforest Observatory Network and the Amazon Rainforest Inventory Network. Over 300,000 trees are being tracked, with more than 1 million diameter measurements in seventeen countries, and the data is standardized and managed by the University of Leeds (at ForestPlots.net).

The concern about a possible Amazon rainforest dieback came from a seminal paper published in 2000 by colleagues at the UK Meteorological Office's Hadley Centre. Their climate model was the first to include vegetation–climate feedback and suggests that global warming by 2050 could increase the winter dry season in Amazonia. For the Amazon rainforest to survive, it requires not only a large amount of rain during the wet season but a relatively short dry season so as not to dry out. According to the Hadley Centre model, climate change could cause the global climate to shift towards a more El Niño-like state with a much longer South American dry season. Kim Stanley Robinson, in his novel *Forty Signs of Rain*, uses the term 'Hyperniño' to refer to a new climate state. The Amazon rainforest could not survive this longer dry season and would be replaced by savannah (dry grassland), which is found both to the east and south of the Amazon basin today. This replacement would occur because the extended dry periods would lead to forest fires destroying large parts of the rainforest. This is exactly what was seen during the two extreme Amazon droughts of 2005 and 2010. The wildfires also return the carbon stored in the rainforest back into the atmosphere, accelerating climate change. The savannah would then take over those burnt areas, as it is adapted to coping with the long dry season, but savannah has a much lower carbon storage potential per square kilometre than rainforest does.

Modelling the Amazon forest response to climate change is complicated because there are positive and negative feedbacks. For example, higher levels of atmospheric CO_2 have a 'fertilization' effect on plants and trees, boosting photosynthesis and promoting growth. They also make plants more water efficient and hence more drought tolerant, offsetting some of the effects of the longer predicted dry season. Other climate models have not found such a profound dieback and the current IPCC review suggests a sustained dieback of the Amazon rainforest is unlikely this century—if the Amazon rainforest stays intact. It is that last part which is the biggest issue, because, under the leadership of Brazilian President Jair Bolsonaro, deforestation rates have been on the rise accompanied by a significant increase in forest fires, many occurring in areas which do not usually suffer from them, indicating that many are being started deliberately. The deforestation and fragmentation of the Amazon and other rainforests around the world make them more vulnerable to climate change and hence make the likelihood of a catastrophic dieback more likely.

Human-induced climate change has already affected our planet and could have even more radical impact over the next 80 years. In addition, scientists worry constantly about potential surprises in the global climate system that could exacerbate future climate change. As discussed above, these include the possibility that Greenland and/or the Antarctic could start to melt irreversibly, raising sea level by many metres in the next century. The North-Atlantic-driven deep-ocean circulation could change, producing extreme seasonal weather in Europe. The Amazon rainforest could start to die back due to the combined effects of deforestation and climate change, causing the loss of huge amounts of biodiversity and increasing carbon emissions to the atmosphere, driving further global warming. Finally, there is the threat of additional CH_4 being released from gas hydrates beneath the oceans and permafrost, which could accelerate climate change. One way to ensure we avoid the worst effects of climate change and greatly reduce the likelihood of climate surprise is to keep climate change as small as possible. The aspiration of our global leaders is to try to keep climate change to just 1.5°C above pre-industrial levels. In Chapter 7, we explore how they came to this decision and how they hope to achieve it.

Chapter 7
Politics of climate change

Introduction

The most logical approach to the climate change problem is to significantly cut GHG emissions. At the Paris climate meeting in 2015, world leaders agreed that global temperature increase should be kept below 2°C, with an aspirational target of 1.5°C. Despite this agreement global carbon emissions have continued to rise every year. The only exception was 2020 when the global lockdown prompted by the Covid-19 pandemic dropped emissions by about 7%. Ceasing almost all flying and car journeys around the world had a small impact on our total GHG pollution. In fact global carbon emissions for 2020 with a global pandemic were the same as 2006. This was because there was very little change to energy production during the pandemic; but there have been calls around the world from business and civil society that the post-pandemic recovery must be a low-carbon one.

Climate change negotiations

The UNFCCC was created at the Rio Earth Summit in 1992 to negotiate a worldwide agreement for reducing GHGs and limit the impact of climate change. The UNFCCC officially came into force on 21 March 1994. As of March 2014, it has 196 parties.

Enshrined within the UNFCCC are a number of principles, including agreement by consensus of all parties, and differential responsibilities. The latter is because the UNFCCC acknowledges that different countries have emitted different amounts of GHGs and therefore need to make greater or lesser efforts to reduce their emissions (Figure 33). For example, in the USA each person on average emits ten times more CO_2 than a person in India. To represent this formally at the negotiations two different groups of parties have been recognized: Annex I countries, which include all the developed countries; and non-Annex I, which include the less developed and rapidly developing countries. Annex I was subsequently divided when some countries argued that their economies were in transition. As a result, the richest countries were placed in an additional category, Annex II. The UNFCCC pays heed to the principle of contraction and convergence: the idea is that every country must reduce its emissions and that all countries must converge on net zero emissions. The net zero emission target emerged from the important IPCC 1.5°C global warming report published in 2018, which clearly showed that to achieve 1.5°C there had to be zero carbon emissions by about 2050 and then negative carbon emissions for the rest of the century.

Kyoto 1997

Since the UNFCCC was set up, the nations of the world, 'the parties', have been meeting annually at the Conference of the Parties (COP) to move negotiations forward. Only five years after the UNFCCC was created at COP3 on 13 December 1997, the first international agreement, the Kyoto Protocol, was drawn up. This stated the general principles for a worldwide treaty on cutting GHG emissions and, more specifically, that all developed nations would aim to cut their emissions by 5.2% on their 1990 levels by 2008–12. The Kyoto Protocol was ratified and signed in Bonn on

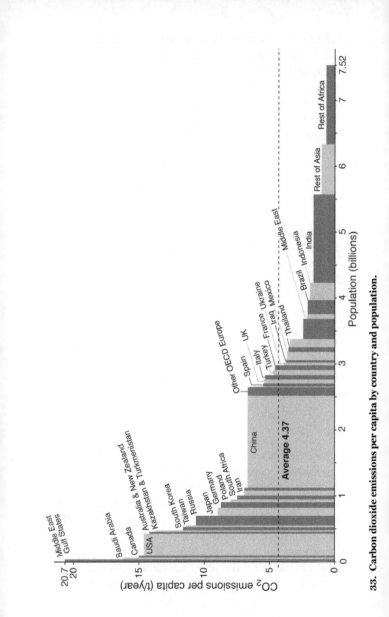

33. **Carbon dioxide emissions per capita by country and population.**

23 July 2001, making it a legal treaty. The USA, under the leadership of President Bush, withdrew from the climate negotiations in March 2001 and so did not sign the Kyoto Protocol at the Bonn meeting. With the USA producing about one-quarter of the world's CO_2 pollution, this was a big blow for the treaty. Moreover, the targets set by the Kyoto Protocol were reduced during the Bonn meeting to make sure that Japan, Canada, and Australia would join. Australia finally made the Kyoto Protocol legally binding in December 2007.

The treaty did not include developing countries. This was to balance out the historic legacy of emissions by developed countries. It was then assumed that developing countries would join the post-2012 agreement. The Kyoto Protocol came into force on 16 February 2005, after Russia ratified the treaty, thereby meeting the requirement that at least fifty-five countries, representing more than 55% of the global emissions, should have signed.

Copenhagen 2009

There were huge expectations of COP15 (Copenhagen) in 2009, despite coming a year after the global financial crash. New quantitative commitments were expected to ensure a post-2012 agreement to seamlessly move on from the Kyoto Protocol. Barack Obama had just become president of the USA. The EU had prepared an unconditional 20% reduction of emissions by 2020 on a 1990 baseline and a conditional target rising to 30% if other developed countries adopted binding targets. Most other developed countries had something to offer. Norway was willing to reduce emissions by 40% and Japan by 25% from a 1990 baseline. Even the USA offered a 17% reduction on a 2005 baseline, which was an equivalent drop of 4% on a 1990 baseline. But the Copenhagen conference went horribly wrong. First the Danish government had completely underestimated the interest in the conference and provided a venue that was too small. So in the second week, when all the high-powered country ministers and their support arrived, there was not enough room, meaning that many NGOs were denied access to the negotiations. Second, it was clear that the negotiators were not ready for the arrival of the ministers and that there was no agreement. This led to the leaking of the 'The Danish Text', subtitled 'The Copenhagen Agreement', and the proposed measures to keep average global temperature rise to 2°C above pre-industrial levels. It started an argument between developed and developing nations as it was brand new text that had just appeared in the middle of the conference. Developing countries accused the developed countries of working behind closed doors and making an agreement that suited them without seeking consent from the developing nations. Lumumba Stanislaus Di-Aping, chairman of the G77, said, 'It's an incredibly imbalanced text intended to subvert, absolutely and completely, 2 years of negotiations. It does not recognize the proposals and the voice of developing countries.'

The final blow to getting an agreement on binding targets came from the USA. Barack Obama arrived only two days before the end of the conference; he convened a meeting between the USA and the BASIC countries (Brazil, South Africa, India, and China), excluding other UN nations, and he created the Copenhagen Accord. The Copenhagen Accord recognized the scientific case for keeping temperature rises below 2°C, but did not contain a baseline for this target, nor the commitments to reduce emissions that would be necessary to achieve the target. Earlier proposals that would have aimed to limit temperature rises to 1.5°C and cut CO_2 emissions by 80% by 2050 were dropped. The agreement made was non-binding and countries had until January 2010 to provide their own voluntary targets. It was also made clear that any country that signed up to the Copenhagen Accord was also stepping out of the Kyoto Protocol. Hence the USA was able to move away from the binding targets of the Kyoto Protocol, which should have been enforced until 2012, and fostered a weak voluntary commitment approach. The Bolivian delegation summed up the way the Copenhagen Accord was reached: 'anti-democratic, anti-transparent and unacceptable'. The legal status of the Copenhagen Accord was also unclear, since it was only 'noted' by the parties, not agreed, as only 122, subsequently rising to 139 countries, agreed to it.

Trust in the UNFCCC negotiations took another blow when in January 2014 it was revealed that the US government negotiators had information during the conference obtained by eavesdropping on meetings of other conference delegations. Documents leaked by Edward Snowden showed how the US National Security Agency (NSA) had monitored communications between countries before and during the conference. The leaked documents show that the NSA provided US delegates with advance details of the Danish plan to 'rescue' the talks should they founder, and also of China's efforts before the conference to coordinate its position with that of India.

Paris 2015

The failure of COP15 in Copenhagen, and its voluntary commitments, cast a long shadow over the successive COP meetings, further darkened by the revelation by Wikileaks that US aid funding to Bolivia and Ecuador was reduced because of their opposition to the Copenhagen Accord. It took over five years for the negotiations to recover from the mess created by Barack Obama and the US negotiators. At COP16 in Cancun and COP17 in Durban, the UNFCCC negotiations were slowly put back on track with the aim of obtaining legally binding targets. Significant progress was made in the REDD+ (Reduced Emissions from Deforestation and Forest Degradation, including safeguards for local people), which is discussed later in this chapter. It was, however, at COP18 in Doha in December 2012 that a second commitment period starting on 1 January 2013 was agreed, which was to last eight years. This ensured that all Kyoto mechanisms

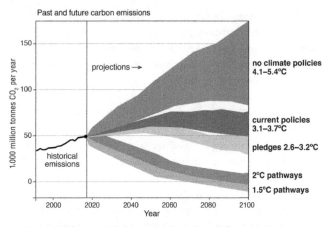

Past and future carbon emissions

← projections →

no climate policies
4.1–5.4°C

current policies
3.1–3.7°C

pledges 2.6–3.2°C

2°C pathways

1.5°C pathways

historical
emissions

1,000 million tonnes CO₂ per year

Year

34. Potential future global warming based on different carbon emissions.

and accounting rules remained intact for this period and parties could review their commitments with a view to increasing them. All this laid the foundations for a future global climate agreement, which was achieved at COP21 in Paris in 2015 (Figure 34).

The climate negotiations in Paris 2015 were a huge success primarily because the French hosts understood the grand game of international negotiation and used every trick in the book to get countries to work together to achieve an agreement signed by all. The agreement states that the parties are required to hold temperatures to 'well below 2°C above pre-industrial levels and to pursue efforts to limit the temperature increase to 1.5°C above pre-industrial levels'. The Paris conference was a high-stakes game of geopolitical poker. Surprisingly, the least powerful countries did much better than expected. The climate talks were subject to a series of shifting alliances going beyond the usual income-rich Northern countries and income-poor countries of the Global South. Central to this was first the US–Chinese diplomacy, as both agreed to limit emissions. Second, a new grouping of countries called the Climate Vulnerable Forum forced the 1.5°C target higher up the political agenda, so much so that it is mentioned in the key aims of the agreement. Political support from the Paris Agreement allowed the IPCC to write the seminal 1.5°C global warming report, which was published in 2018. This report documented the significant increase in impact between a 1.5°C and a 2.0°C world. It also documented how a 1.5°C world could be achieved—which, as noted above, requires net zero carbon emissions by 2050 and then for carbon to be taken out of the atmosphere for the rest of the century (Figure 35). The quicker the world gets to net zero, the less carbon needs to be extracted from the atmosphere between 2050 and 2100. The Paris Agreement was just the start of the process, because taking into account all the country pledges and assuming they will be fulfilled, the world would still warm by about 3°C (Figure 34).

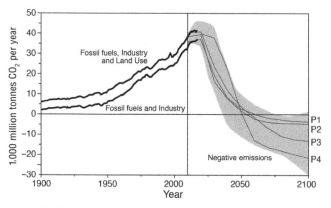

35. Achieving a 1.5°C world.

In 2017, the Paris Agreement had a major setback. President Trump declared he was taking the USA out of the agreement, as he believed it was unfair and biased towards developing countries. In accordance with Article 28 of the Paris Agreement, a country cannot give notice of withdrawal from the agreement before three years of its start date in the relevant country. So, the earliest possible effective withdrawal date by the USA was 4 November 2020—one day after the 2020 US presidential election. One of the first acts of the newly elected President Biden was to re-engage the USA in the Paris Agreement.

The new president faces additional challenges because over the four years of the Trump presidency nearly a hundred environmental rules and regulations have been rescinded or are in the process of being removed. These included rolling back the Obama administration's fuel efficiency and emission standards for vehicles; reductions in their coal-emission standards for coal-fired power plants; and the weakening of the efficient lighting regulation, meaning less efficient light bulbs could still be purchased after 2020.

President Trump approved two controversial oil pipelines (Keystone XL and Dakota Access), allowed drilling in nearly all US waters, opened the Arctic National Wildlife Refuge to drilling, creating a huge expansion of oil and gas exploration. In 2021 President Biden rescinded all these executive orders, re-engaged the US in the Paris Agreement, invested heavily in low carbon technology and infrastructure and pledged to reduce the US carbon emissions by 50% by 2030 and net zero carbon by 2050.

Glasgow 2021 and Sharm El Sheikh 2022

In 2021, the UK and Italy co-hosted COP26 in Glasgow, after a year's delay due to the COVID-19 pandemic. This was the first global stocktake following the Paris Agreement, with countries submitting Nationally Determined Contributions (NDCs) or their pledges to cut greenhouse gas emissions. The NDCs submitted show that over 90% of the world's GDP now sits under Net Zero emissions targets. If all the NDCs are fulfilled, then the global temperature rise could be kept to between 2.4°C and 2.7°C. This is nowhere near the Paris Agreement 1.5°C target that was re-emphasised in the Glasgow Climate Pact signed by all 197 countries. So countries were asked to submit new, more ambitious, NDCs for COP27 in Egypt in 2022 – breaking the 5-year cycle of the Paris Agreement. However these failed to materialise and a new call was made for COP28 in the UEA. The Glasgow Climate Pact also called for the phasing down of coal and the removal of inefficient fossil fuel subsidies – the first time that fossil fuels have been mentioned in any international climate treaty. COP26 succeeded in finishing Article 6 – the rules and regulations governing monitoring and trading of carbon emissions and sinks between countries and other organisations. In 2022 little progress was made at COP27 in Sharm El Sheikh. An agreement to set up a 'Loss and Damage' facility was achieved – but who pays and who can claim will have to be agreed at future meetings. The 2010 promise of $100 billion per year from developed to developing countries to help rapid decarbonisation still has not materialised.

Is the UNFCCC process flawed?

Various flaws have been pointed out in the approach of the UNFCCC. Here are some of the major ones.

Not going far enough. The first perceived flaw in the UNFCCC procedure is that despite 25 years of negotiations it has failed to deliver any lasting agreement. As mentioned above, the current Paris Agreement, if all the pledges are honoured, still causes global warming of at least 3 °C (Figure 34) and the associated impacts described in Table 4.

No enforcement. The fundamental problem with international agreements and treaties is that there are no real means of enforcement. This was one of the arguments that the USA used when proposing the Copenhagen Accord, suggesting that even binding targets must in effect be voluntary as countries decide whether or not to comply. This is why policies and laws are required at a regional level, as in the EU, and at a national level, such as in the UK with its Climate Change Act. The only way to translate international treaties into a reality is through regional and national laws. This multi-level governance is also required to stop gaming of particular systems.

Green colonialism. Many social and political scientists have raised philosophical and ethical doubts about climate negotiations as a whole. The main concern is that they reflect a version of colonialism, since rich developed countries are seen to be dictating to poorer countries how and when they should develop. For many years countries such as India and China have resisted calls to cut their emissions, stating that it would damage their development and attempts to alleviate poverty. Others have supported measures such the Clean Development Mechanism (CDM), which allows developed countries to pay for emission reductions in a developing country so that it counts towards their national reduction target. They also provide a development dividend, moving money from the rich to the poorer countries. But 80% of the CDM project credits were allocated to China, Brazil, India, and Korea, the richest developing countries, so funding did not reach the world's poorest. Also, 60% of the CDM carbon credits have been purchased by the UK and the Netherlands, resulting in a very skewed financial exchange.

Nation versus sector approach. The UNFCCC approach has another problem, which is embedded in the concept of the nation-state and is a major issue in a global capitalist world with supposedly free trade. For example, if the USA through the Paris Agreement wanted to reduce carbon emissions from heavy industry, it could impose a carbon tax on steel and concrete production. If other countries in the world do not have this restriction, their products become cheaper, even including the cost of transportation by ship, air, or road to the USA, all of which would lead to the emission of more CO_2 overall. So global economics can undermine any national attempts to do the right thing and reduce their emissions. An alternative approach would be for global agreements to be made at a sector level. For example, there could be a global agreement on how much carbon can be emitted per ton of steel or concrete produced. All countries could then agree only to buy steel or concrete produced in this low-emission way, which would make for a fairer trading scheme, with countries not losing out as a result of changes within their industries to lower GHG emissions. The problems are, of course, how to police such a scheme across so many different industrial sectors.

Carbon trading

Many politicians have advocated using either regional or global carbon trading schemes. The most successful system of carbon trading is 'cap and trade', whereby politicians set a cap, a maximum total of pollution allowed, and a trading system is then set up so that different industries can trade credits. It is acknowledged that the various industries can clean up at varying rates and costs, and this trading system allows the most cost-effective approach to be found. This type of system was successful in the USA in reducing air pollution by trading sulfur dioxide and N_2O emissions. The US Clean Air Act of 1990 required electrical utilities to lower their emissions of these pollutants by 8.5 million tonnes compared with 1980 levels. Initial estimates in 1989 suggested it would cost $7.4 billion; a report in 1998 based on actual compliance data suggested it had cost less than $1 billion.

Currently over 13% of global carbon emissions are covered by national or regional carbon trading schemes. These include schemes in the USA, Canada, China, South Korea, Japan, Brazil, Argentina, South Africa, and the EU. The EU's Emissions Trading Scheme (ETS) is the largest and longest running carbon trading scheme. It covers more than 11,000 installations with a net energy use of 20 megawatts, and includes electricity generation, ferrous metal production, cement production, refineries, pulp, paper, and glass manufacturing. The ETS covers thirty-one countries, consisting of all twenty-eight EU member states plus Iceland, Norway, and Liechtenstein. The ETS covers half the EU's CO_2 emissions and 40% of its total GHG emissions. Under the 'cap and trade' principle, a cap is set on the total amount of GHGs that can be emitted by installations in each country. 'Allowances' for emissions are then auctioned off or allocated for free, and can subsequently be traded. Installations must monitor and report their CO_2 emissions, ensuring they hand in enough allowances to the authorities to cover their emissions. If emissions exceed what is permitted by its allowances, an installation must purchase allowances from others. Conversely, if an installation has performed well at reducing its emissions, it can sell its leftover credits. This allows the system to find the most cost-effective ways of reducing emissions without significant government intervention. The ETS has been arranged in four phases, 2005–7, 2008–12, 2013–20, and 2021–30. In each phase the total number of credits available has been reduced and the number of sectors and industries included has been increased; this ratchet approach has been used to drive emissions down as quickly as possible. In 2020, it was estimated that the EU ETS had reduced CO_2 emissions by more than 1 billion tonnes between 2008 and 2016, in other words 3.8% of total EU-wide emissions. The EU ETS has, however, been criticized because the emission caps have not been strict enough, leading to a very low-carbon price. In the UK, the addition of a 'carbon price floor', or a minimum government-set carbon price, has been essential in removing coal from the energy mix.

REDD+

The idea of developing an instrument on deforestation within the climate change negotiations was first suggested at COP11 (2005) in Montreal and was referred to as RED (Reduced Emissions from Deforestation). The UN REDD (Reducing Emissions from Deforestation and Forest Degradation) programme was agreed in principle at COP13 (2007) in Bali. This has been subsequently refined as REDD+, the '+' representing safeguards to protect local people and safeguards to the local ecosystem and biodiversity. REDD+ (or REDD-plus) is visualized as a win-win solution that can protect forests and ecosystems, promote reforestation, and protect and compensate forest dwellers from and for lost income from exploiting their forested land. Each REDD+ project has to be submitted for verification to ensure it does provide win-win outcomes before it can go ahead and be funded.

At COP19 in 2013 REDD was developed further and the 'Warsaw Framework for REDD-plus' was agreed which included how to monitor, measure, report, and verify forest changes and credits. The remaining outstanding decisions on REDD+ were completed at COP21 in Paris in 2015, including how to report on the safeguards using non-market approaches, and how to account for non-carbon benefits. So by 2015 the UNFCCC rulebook on REDD+ was completed. All countries have been encouraged to implement and support REDD+, which is in Article 5 of the Paris Agreement. This was part of the broader article that specified that all countries should take action to protect and enhance their GHG sinks and stores such as forests.

In 2009, US intransigence and the impact of the global financial crash meant that the Copenhagen climate negotiation failed to agree on any successor to the Kyoto Protocol. It took six years to get the climate negotiations back on track with the success of COP21 and the signing of the Paris Agreement. In 2020, COP26 in Glasgow was delayed by the Covid-19 pandemic, but unlike with the global financial crash, the global pandemic did not derail the climate negotiation. Instead, around the world, more and more companies, organizations, and individuals called for a better, healthier, and safer world after the pandemic. This was because the world saw that there could be a different relationship between government, industry, and civil society—a relationship where the health and wellbeing of citizens is put before the economic gains of a country or small minority of individuals. In addition, the new narrative of global 'net zero carbon' emissions by 2050 is very powerful—it changes the discussion from how much we can reduce emissions to when we will get rid of them altogether. The challenge is huge because in less than 30 years we have to shift from emitting over 40 billion tonnes of CO_2 per year to zero (Figure 36). The Paris Agreement pulls no punches and makes it

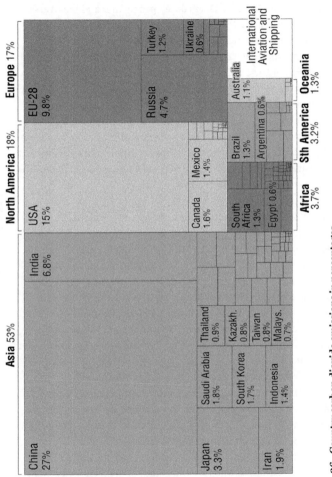

Asia 53%

China 27%

Japan 3.3%

Saudi Arabia 1.8%

South Korea 1.7%

Indonesia 1.4%

Iran 1.9%

Thailand 0.9%

Kazakh. 0.8%

Taiwan 0.8%

Malays. 0.7%

India 6.8%

North America 18%

USA 15%

Canada 1.6%

Mexico 1.4%

Europe 17%

EU-28 9.8%

Russia 4.7%

Turkey 1.2%

Ukraine 0.6%

International Aviation and Shipping

Oceania 1.3%

Australia 1.1%

Sth America 3.2%

Brazil 1.3%

Argentina 0.6%

Africa 3.7%

South Africa 1.3%

Egypt 0.6%

36. **Country carbon dioxide emissions in percentages.**

clear that if we are to stabilize climate change even at 2 °C then we require the complete transformation of energy generation, industry, infrastructure, and individual behaviours around the world. In fact, we need to use every single solution we have to tackle climate change.

Chapter 8
Solutions

Introduction

There are three types of solutions to climate change. The first is adaptation, which is providing protection for the population from the impacts of climate change. The second is mitigation, which in its simplest terms is reducing our carbon footprint and thus reversing the trend of ever-increasing GHG emissions. Third is geoengineering, which involves large-scale extraction of CO_2 from the atmosphere or modification of the global climate.

Adaptation

There have already been climate change impacts, and these will increase as the global temperature continues to rise. The second report of the IPCC Sixth Assessment examines the impacts of climate change and the potential sensitivity, adaptability, and vulnerability of each national environment and socioeconomic system. The IPCC gives six clear reasons why we must adapt to climate change: (1) climate change impacts cannot be avoided even if emissions are cut rapidly to zero (see Chapter 4); (2) anticipatory and precautionary adaptation is more effective and less costly than forced last-minute emergency fixes; (3) climate change may be more rapid and more pronounced than current estimates suggest, and unexpected and extreme events are likely to occur; (4) immediate benefits can be gained from better adaptation to climate variability and extreme atmospheric events (e.g. with the storm risks, strict building laws and better evacuation practices would need to be implemented); (5) immediate benefits can also be gained by removing maladaptive policies and practices (e.g. building on floodplains and vulnerable coastlines); and (6) climate change brings opportunities as well as threats. Figure 37 provides an example of how countries can adapt to the predicted sea-level rise.

The major threat from climate change is its unpredictability (see Chapter 6). As noted earlier, humans can live within a huge range of climates, from baking deserts to the frozen Arctic, but we have only been able to do so because we have been able to predict the extremes of weather we must cope with. As climate change impacts increase then the weather will become both more extreme and more unpredictable. So both physical and social adaptations are required to protect people's lives and livelihoods.

Physical adaptations require us to think about how to change our infrastructure. For example, will we need to build better sea defences, more reservoirs, restore wetlands, or retrofit buildings with air conditioning? In many countries, these large infrastructure projects can take up to 30 years to plan, develop, and build. If we take the issue of sea-level rise (Figure 38), it can take 10 years to research and plan appropriate measures to cope with it. It can then take another 10 years for the full consultative and legal processes; and a further 10 years to implement these changes. It can take another decade for the natural restoration to take place to complete the sea-level adaptation project (Figure 38). A good example of this is the Thames Barrier that currently protects London from flooding: it was developed in response to the severe flooding in 1953 but did not open officially until 1984, 31 years later.

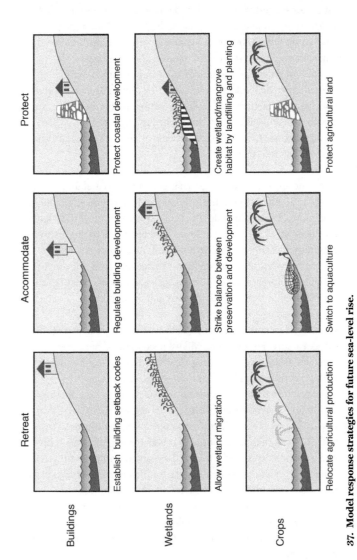

Retreat   **Accommodate**   **Protect**

Buildings

Establish building setback codes   Regulate building development   Protect coastal development

Wetlands

Allow wetland migration   Strike balance between preservation and development   Create wetland/mangrove habitat by landfilling and planting

Crops

Relocate agricultural production   Switch to aquaculture   Protect agricultural land

37. Model response strategies for future sea-level rise.

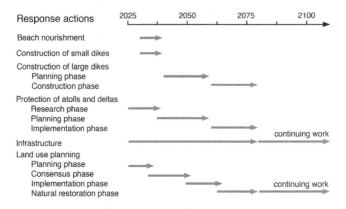

38. Lead times for response strategies to combat climate change.

We must also consider social adaptations and changes in people's behaviour. After the 2003 European heatwave, France completely reassessed its health response to the crisis. They changed everything, including: communication with the public; vulnerable individuals' health checks; local health responses; and hospital admissions and treatment. It is estimated that in subsequent heatwaves the death toll was cut by over 75% because of these social adaptations. In many countries food and water security will be the major issue, and therefore policies to safeguard people's access to food and clean water even when they are unable to pay for it will be essential. In many ways the most important adaptation to climate change is good governance, so that policies can be formulated and enacted to protect the most vulnerable people in society.

There are, however, limits to adaptation. In some regions, climate change impacts may become so great that they go beyond our ability or finances to protect the population living there. Continued sea-level rise will mean many small island nations may become uninhabitable and the population will have to relocate. In 2019, Indonesia's President Joko Widodo announced that the national capital would move from Jakarta, on the island of Java, to the province of East Kalimantan, on Borneo. This was in part to relieve pressure on the capital and address inequality in Indonesia, but it was also because Jakarta is sinking. Areas of north Jakarta, even with the sea wall designed to protect the population, are falling at an estimated 25 cm a year due to subsidence. This is due to sea-level rise and the extraction of fresh water from shallow aquifers, leading to subsidence.

The other problem is that adaptation requires money to be invested now; but many countries just do not have that money and, where they might raise it, their citizens are unwilling to pay more taxes towards protecting themselves in the future, since most people live for the present. Many countries have short election cycles of four to five years, which means politicians are always thinking about the short term and rarely about the longer term, and this limits their insight into and investment in adaptation projects. This is, of course, despite the fact that all of the adaptations discussed will in the long term save money for the local area, the country, and the world.

Mitigation

The requirement to cut global carbon emissions by half by 2030 and hit net zero by 2050 is extremely challenging, and will mean using every available solution as soon as possible. There is some good news. GDP growth and carbon emissions have over the past decade become completely decoupled, with a large rise in world GDP compared with a much smaller rise in carbon emissions (see Figure 39). What we need to do now is to invert the relationship so that as GDP rises, carbon emissions drop. In 2020, the International Energy Agency (IEA) and International Monetary Fund (IMF) published a report recommending massive investment into clean energy, which would create millions of new jobs. It was realized during the Covid-19 pandemic, that energy generation and use were the key to dropping carbon emissions.

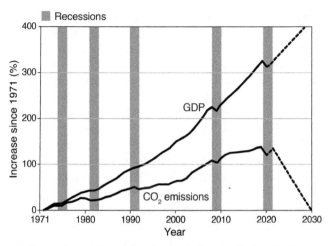

39. Comparison of GDP and carbon-emission growth since 1971.

The report outlines plans for mass home renovations, fossil-fuel subsidy reforms, expansion of renewable energy, and power grids. Several of these are discussed below.

Alternative, renewable, or clean energy

Fossil fuels were an amazing discovery, and they have allowed the world to develop at a faster rate than it has at any other time in history. The high standard of living in the developed world is based on cheap and relatively safe fossil fuels. But burning fossil fuels has had the unintended consequence of changing global climate. So in the 21st century we need to switch from fossil-fuel energy to low-carbon or carbon-neutral energy. This includes solar, bio, wind, hydro, wave, and tidal energy.

Solar power. The Earth receives on average 343 W/m^2 from the Sun, and yet as a whole the planet only receives two-billionth of all the energy put out by the Sun. The Sun is the ultimate source of energy, which plants have been utilizing for billions of years. At the moment, we can convert solar energy directly to heat or electricity, and we can capture the energy through photosynthesis by growing biofuels. The simplest approach is through solar heating. On a small scale, houses and other buildings in sunny countries can have solar heating panels on the roof, which heat up water, so people can have carbon-free hot showers and baths. On a large scale, parabolic mirrors are used to focus the solar energy to generate hot liquid (water or oil) to drive turbines to create electricity. The best places to situate solar heat plants are in low-latitude deserts, which have very few cloudy days per year. Solar heat plants have been built in California since the 1980s and are now being built in many other countries. Solar photovoltaic panels convert sunlight directly into electricity. The individual rays of the Sun hit the solar panel and dislodge electrons inside it, creating an electrical current. The main advantage of solar panels is that you can place them where the energy is needed and avoid the complicated infrastructure normally required to move electricity around. Over the past decade there has been a massive increase in their efficiency, the best commercially available solar panels being about 23% efficient, which is significantly more than photosynthesis at about 1%. Solar panel efficiency goes up during winters, as they work better in cold temperatures, although, of course, then they also produce less electricity due to the shorter days and less intense sunlight. There has also been a significant drop in price due to huge investment in the technology.

Biofuels. These are the product of solar energy converted into plant biomass via photosynthesis, which can then be used to produce liquid or solid fuels. The global economy is based on the use of liquid fossil fuels, particularly in the transport sector. So, in the short term, fuels derived from plants could be an intermediate low-carbon way of powering cars, ships, and aeroplanes. Ultimately, electric cars are the future, because the required electricity can be produced carbon-neutrally. However, this energy source is not an option for aeroplanes. Traditional air fuel, 'kerosene', combines relatively low weight with high energy output. Research is being carried out to see whether a biofuel can be produced that is light enough and powerful enough to replace kerosene. Many power stations around the world have been converted so that they can burn wood pellets instead of coal or natural gas to create steam to turn the turbines to make electricity. Critics have argued that wood pellets are not sustainable and do not have as low a carbon footprint as claimed.

Wind power. Wind turbines are an efficient means of generating electricity, if they are large and preferably located out at sea. Ideally, we need turbines the size of the Statue of Liberty for maximum effectiveness. The London Array has been built in the River Thames estuary, 12 miles from the Kent coast, and consists of 175 turbines. It will generate over 2,500 megawatts (MW), making it the world's largest offshore wind farm. It can power up to half a million homes and reduce harmful CO_2 emissions by nearly 1 million tonnes per year. There are some problems with wind turbines. First, they do not supply a constant source of electricity; if the wind does not blow or it blows too hard, no electricity is generated. Second, people do not like them, finding them ugly, noisy, and a worry in terms of the effects they may have on local natural habitats. All these problems are easy to overcome by situating wind farms in remote locations, out at sea, and away from areas of special scientific or natural interest. Recent research has shown little or no effect on local wildlife even when wind turbines are situated close to land. One study suggests that wind in principle could globally generate over 125,000 terawatt-hours, which is five times the current global electricity requirement.

Wave and tidal power. Wave and tidal power could also be an important source of energy in the future. The concept is simple: to convert the continuous movement of the ocean in the form of waves into electricity. However, this is easier said than done, and experts in the field suggest that wave power technology is 20 years behind solar panel technology. Tidal power has one key advantage over solar and wind power—it is constant. In any country, for energy supply to be maintained at a constant level, there has to be at least 20% production guaranteed, known as the baseline requirement. With the switch to alternative energy, new sources of energy to provide this consistent baseline need to be developed.

Hydroenergy. Hydroelectric power is globally an important source of energy: in 2010, it supplied 5% of the world's energy. The majority of the electricity comes from large dam projects. These projects can present major ethical problems as large areas of land must be flooded above the dam, requiring the mass relocation of people and destruction of the local environment. A dam also slows water flowing down a river and prevents nutrient-rich silt from being deposited lower down. If the river crosses national boundaries, there are potential issues over the rights to water and silt. For example, one of the reasons why Bangladesh is sinking is the lack of silts due to the dams built on the major rivers in India. There is also debate about how much GHG emissions hydroelectric plants save, because even though the production of electricity does not cause any carbon emissions, the rotting vegetation in the area flooded behind the dam does give off significant amounts of CH_4.

Geothermal energy. Below our feet, deep within the Earth, is hot molten rock. In some locations, for example in Iceland and Kenya, this hot rock comes very close to the Earth's surface and can be used to heat water to make steam. This is an excellent carbon-free source of energy, because part of the electricity generated from the steam is used to pump the water down to the hot rocks. Unfortunately, its availability is limited by geography. There is, however, another way the warmth of the Earth can be used. All new buildings could have a borehole below them with ground-sourced heat pumps. Cold water can then be pumped down into these boreholes, with the ground warming the water up, cutting the cost of providing hot water to buildings. This method could be used almost everywhere in the world.

Nuclear fission. Energy is generated when heavy atoms such as uranium are split—a process known as nuclear fission. The process has a very low direct carbon signature, but a significant amount of carbon is generated in mining the uranium, building the nuclear power station, decommissioning the power station, and safely storing and disposing of nuclear waste. At the moment, 5% of global energy is generated by nuclear power. The new generation of nuclear power stations are extremely efficient, achieving nearly 90% of the theoretically possible energy production. The main disadvantages of nuclear power are the generation of high-level radioactive waste and concerns about safety, although improvements in efficiency have reduced the waste levels and the new generations of nuclear reactors have state-of-the-art safety features built in. The 1986 Chernobyl disaster and the 2011 Fukushima Daiichi nuclear disaster illustrate that nuclear plants are, however, still not safe, being vulnerable to human error and natural hazards. The advantages of nuclear power, however, are that it is reliable and can provide the required base load in the energy mix, and the technology for its use is already available and thoroughly tested.

Nuclear fusion. This process involves the generation of energy when the nuclei of two small atoms fuse together. It is the process that lights up our Sun and every other star. The idea is that the heavy form of hydrogen found in seawater, deuterium, can be combined with the other heavy isotope of hydrogen, tritium; and the only waste product is the non-radioactive gas, helium. The problem, of course, is persuading those nuclei to fuse. In the Sun, fusion occurs in the core, at incredibly high temperatures and pressures. Great advances in fusion technology have been made around the world but what is required now is huge investment to make fusion commercially viable.

Carbon capture and storage

Removal of CO_2 during industrial processes can be tricky and costly, because not only does the gas need to be removed, but it must be stored somewhere as well. The IPCC Special Report on Carbon Dioxide Capture and Storage published in 2005 concluded that the technology for carbon capture and storage (CCS) existed but that there is little commercial experience in configuring all of the components needed to create fully integrated CCS systems at the kinds of scales needed in the future. Power production costs that include CCS would rise by at least 15% and could be as high as 100%. Not all the recovered CO_2 has to be stored; some may be utilized in enhanced oil recovery, the food industry, chemical manufacturing (producing soda ash, urea, and methanol), and the metal-processing industries. CO_2 can also be applied to the production of construction material, solvents, cleaning compounds, and packaging, and in waste-water treatment. In reality, most of the CO_2 captured from industrial processes would have to be stored. It has been estimated that theoretically two-thirds of the CO_2 formed from the combustion of the world's total oil and gas reserves could be stored in corresponding reservoirs. Other estimates indicate storage of 90–400 Gt in natural gas fields alone and another 90 Gt in aquifers.

Oceans could also be used to dispose of the CO_2. Suggestions have included storage by hydrate dumping: mixing CO_2 and water at high pressure and low temperatures creates a solid, or hydrate, which is heavier than the surrounding water and thus drops to the bottom (see Figure 32). Another more recent suggestion is to inject the CO_2 half a mile deep into shattered volcanic rocks in between giant lava flows. The CO_2 will react with the water percolating through the rocks. The acidified water will dissolve metals in the rocks, mainly calcium and aluminium. Once it forms calcium bicarbonate (HCO_3^-) with the calcium, it can no longer bubble out and escape. If it does escape into the ocean, then HCO_3^- is relatively harmless. With ocean storage there is the added complication that the oceans circulate, so whatever CO_2 is dumped, some of it will eventually return. Moreover, scientists are very uncertain about the effects of this solution on the ocean ecosystems.

The major problem with all these methods of storage is safety. CO_2 is a very dangerous gas because it is heavier than air and can cause suffocation. This was powerfully illustrated in 1986, when a large release of CO_2 from Lake Nyos, in the west of Cameroon, killed more than 1,700 people and livestock up to 25 km away. Although similar disasters had previously occurred, never had so many people and animals been asphyxiated on such a scale in a single brief event. Scientists now believe that dissolved CO_2 from the nearby volcano had seeped into the lake from springs below and had lain trapped in deep water by the weight of water above. In 1986, there was an avalanche that churned up the lake waters, resulting in an overturn of the whole lake and an explosive release of all the trapped CO_2. Nevertheless, huge amounts of mined ancient CO_2 is pumped around the USA to enhance oil recovery, and there have been no reports of any major incidents. Engineers working on these pipelines feel they are much safer than gas and oil pipelines, many of which run across most major US cities.

Transport

One of the greatest challenges to mitigating GHG emissions is transport. At the moment, transport accounts for 14% of GHG emissions globally. In many developed countries the carbon emissions from energy production, business, and residential sectors are all going down despite annual growth in the economy; but transport emissions, mainly from cars and aviation, are still increasing. Many in the developing world aspire to the same level of car ownership and international travel as the developed world and hence there is the potential for huge growth in transport emissions.

Electric cars, both in terms of range and performance, have improved greatly over the past decade and there is general acceptance that they represent the future. This acceptance has been accelerated by the 2020 pandemic, during which, in many regions, road traffic almost ceased due to lockdowns and everyone noticed the huge improvement in air quality. If there were a switch to 100% electric vehicles, there would be a 50% cut in air pollution. The constant wearing down of tyres, brake pads, and the tarmac on roads also creates air pollution, which accounts for the other 50%. The impact of electric cars on carbon emissions could be significant, but it would rely on there being a guaranteed supply of low-carbon or carbon-neutral electricity. In the UK from 2034 only electric cars will be sold and fossil-fuel engines will be banned by 2040, while in California all new passenger vehicles sold from 2035 onwards must be zero emission.

International shipping and aviation account for 3.2% of global GHG emissions per year. Aeroplanes have become an easy target for climate change campaigners as international flights are a highly visible symbol of consumption and have never been covered by an international treaty. There is a need for incentives to improve the carbon efficiency of flights and ultimately to make them as close to carbon neutral as possible. The fundamental issue is that currently an international treaty prohibits the taxation of aviation fuel. The Convention on International Civil Aviation, also known as the Chicago Convention, was signed in 1944 and has been revised eight times. It deals with the rules and regulations required to allow flights between countries. It also states that fuel, oil, spare parts, regular equipment, and aircraft stores are exempted from any form of taxation, which means a carbon tax on aviation fuel to drive efficiency gains is currently not permitted. This is unfortunate, as not only can we build much more efficient aeroplanes today, but there are alternative fuels that could be used. Biofuels could be developed as an additive or even as a replacement for traditional air fuel kerosene. It is also possible to create artificial kerosene, by extracting CO_2 from the atmosphere and combining it with water. This takes a huge amount of energy, but if electricity from renewable sources were used, it could be possible to have carbon negative aviation fuel. This would still only work if there were regulations or a carbon tax in place to make it cost effective to create artificial kerosene. In the short term, as there is no real fuel solution for aviation, the airlines are keen to be involved in carbon trading. This way, the airlines can 'offset' their carbon emissions by ensuring an equivalent amount is saved elsewhere.

The other alternative is to persuade people to use public transport instead of their car or flying. For most people it is clear that providing cheap and accessible electric buses, taxis, subway systems, and railways would all help reduce the number of car journeys being made. Public transport could also help with freight and goods deliveries, as the railway network could be used at night to transport goods around the country and between countries. Railways could also be used to replace internal and international flights. It has been calculated that all internal flights between American cities less than 600 miles apart could be replaced by high-speed electric 'bullet' trains travelling over 200 miles per hour, providing a quicker, safer, and cleaner way to get around. This would remove 80% of the flights within the USA, but would require high-speed trains running up and down the east and west coasts—with connections to the two major hubs of Chicago and Atlanta. This sort of high-speed train network already exists in Japan, South Korea, and parts of China and the EU, it just needs to be extended to the rest of the world.

The 2020/1 Covid-19 pandemic has boosted the use of internet and video-conferencing, showing a lot of commuting can be avoided as many people are happier working from home. It has also demonstrated that many international meetings, including huge scientific conferences, can be done very successfully using remote access technology. If this leads to a long-term decline in local and international travel then decarbonizing our transport networks will become easier.

Fossil-fuel subsidies

One of the major political problems with reducing carbon emissions concerns energy subsidies. First, there are huge fossil-fuel subsidies, which continue to make oil, gas, and coal relatively cheap. Second, there is resistance to providing subsidies and tax incentives to the energy companies to build and supply renewable energy at competitive rates. A recent report from the International Monetary Fund suggests that the fossil-fuel industry receives over $5.2 trillion per year in subsidies (nearly twice the size of the UK's annual GDP)—this includes direct payments, tax breaks, reduced retail prices, and the cost of climate change damage. Of this governments provide at least $775 billion to $1 trillion as subsidies and at least $444 billion per year in direct funding to oil, gas, and coal companies to support exploration, extraction, and development. There is also a tremendous security cost associated with fossil fuels. A large part of foreign policy and military strategy for many countries involves protecting shipping lines for fossil fuels. The US military spends at least $81 billion a year protecting oil supplies. In comparison there are no aircraft carriers defending wind turbine supply chains or strategic silicon reserves for solar panels.

The IMF report shows that fossil fuels account for 85% of all global subsidies and that they remain a large part of domestic policies. Had nations reduced subsidies in such a way as to create efficient fossil-fuel pricing in 2015, the IMF believes that it 'would have lowered global carbon emissions by 28% and fossil-fuel air pollution deaths by 46%, and increased government revenue by 3.8% of GDP.' It seems fossil-fuel subsidies are bad for the environment and bad for the economy.

So why do fossil-fuel subsidies persist? This may be down to the ownership of the major oil and gas companies. Out of the top 26 oil and gas companies only seven are private companies; the other

19 are fully or partly owned by countries. Hence the state-owned companies are making huge amounts of money for the country and will continue to be given state aid in the form of subsidies and tax breaks to ensure that they are competitive with other oil and gas producing nations. This is only set to get worse with fracking and the shale gas revolution, with many countries such as the USA and the UK having found new reserves of natural gas underground.

Carbon trading, taxation, and offsetting

There are three main policy approaches that can be used to help reduce carbon net emissions. The first way to reduce carbon emissions is to impose a carbon tax on activities and goods that emit a large amount of carbon. Most economists agree that carbon taxes are the most efficient and effective way to curb emissions, with the least adverse effects on the economy. To avoid these taxes being regressive the income should be used to support the least well off in society who will be the worst affected by these taxes. Carbon taxes have been implemented in 25 countries, while 46 countries put some form of price on carbon, either through carbon taxes or emissions trading schemes.

The second approach, as discussed in Chapter 7, is carbon trading, whereby carbon emissions are limited by issuing of carbon permits. Carbon trading can drive innovation and drive down costs. It is also a way of making renewable energy and CCS economically viable. Some emissions trading schemes allow companies to buy carbon offsets either nationally or internationally to count against their total emissions.

A carbon offset is defined as a reduction in emissions of CO_2 or other GHGs made in order to compensate for emissions made elsewhere. This can be either by increased carbon storage through reforestation programmes or by retiring or removing emissions, for example closing down a coal-fired power station. There are two main carbon offsetting systems: the UN CDM and the voluntary markets. The CDM has been described in Chapter 7 and involves UN-certified programmes in developing countries being funded to make significant GHG savings. The voluntary system peaked in volume in 2008 but has seen a substantial increase since 2018. This is because a large number of companies around the world have adopted science-based targets, which means they want to be carbon neutral by 2050 if not earlier. They include fifteen airlines, such as EasyJet, British Airways, and Emirates, which have all announced major carbon offset schemes.

Offsets are an important policy tool and can help reduce total emissions, especially in sectors where it is very difficult at the moment to reduce emissions. New national and international regulations are required to ensure proper monitoring and verification of offsets. Moreover, oversight is required to ensure that companies do not game the system by creating emissions just so that they can be paid to stop them. For example, one Chinese company generated $500 million in carbon offsets by installing a $5 million incinerator to burn the hydrofluorocarbons (HFCs) produced by the manufacture of refrigerants. Many companies followed this approach, and HFCs are no longer allowed under offset schemes.

Reforestation and rewilding

One of the most important ways to remove CO_2 from the atmosphere is through reforestation and rewilding. Since the beginning of agriculture it has been estimated that humans have cut down 3 trillion trees—about half the trees on Earth. So we know that the Earth can sustain a much larger forested area. Rewilding habitats and reforesting may be easier in the future as the world is already becoming a wilder place. This may seem counterintuitive, given that the global population will grow from 7.8 billion today to 10 billion by 2050, but by then nearly 70% of us will live in cities and we will have abandoned many remote rural areas, making them ripe for restoration. Already, in Europe, 2.2 million hectares of forest has regrown per year between 2000 and 2015. In Spain, forest cover has increased from 8% of the country's territory in 1900 to 25% today, while in the UK forest cover hit a low of 5% after the First World War and is now back up to 13%.

Massive reforestation isn't a pipe dream; it can have real benefits for people. In the late 1990s, environmental deterioration in western China became critical, with vast areas resembling the Dust Bowl of the American Midwest in the 1930s. Six bold programmes were introduced, targeting over 100m hectares of land for reforestation. Grain for Green is the largest and best known of these. These radical tree-planting programmes had an amazing effect as the trees stabilized the soils, greatly reducing soil erosion and the impacts of flooding. Through transpiration, the trees added moisture to the atmosphere, reducing evaporation and water loss. Once the forests reached a critical size and area they also started to stabilize the rainfall. All these impacts combined to boost local agricultural production. The ongoing programme has also helped to alleviate poverty as direct payments were made to farmers to set aside their land for reforestation. This was an excellent example of the win-win solution required for climate change as it increased carbon storage, improved the local environment, and helped alleviate extreme poverty.

In 2019 researchers claimed in the journal *Science* that covering 900m hectares of land—roughly the size of the continental US—with 1 trillion trees could store up to 205 billion tonnes of carbon, about two-thirds of the carbon that humans have already put into the atmosphere. The '1 trillion trees' mantra has caught the popular imagination and even President Trump declared at the Davos meeting that this was a good idea. The only problem is that the research that produced this high carbon number was fundamentally flawed and in fact the IPCC and other studies suggest that new forests could store on average an extra 57 billion tonnes of carbon by the end of the century. This is still a large number, but considering we put 11 billion tonnes of carbon into the atmosphere per year, this represents only five years of human pollution. So reforestation is not an alternative to rapid and deep cuts in our fossil-fuel emissions. But later in this century we will require negative carbon emissions to keep the global warming to 1.5°C—one of the key ways of creating negative emissions is through reforestation.

Already 63 countries have joined the Bonn Challenge and pledged to restore 350m hectares of degraded land to forest worldwide. That's an area fifteen times the size of the UK. But there is another issue. Massive reforestation only works if the world's current forest cover is maintained and increased. As noted earlier, deforestation of the Amazon rainforest—the world's largest—has increased since Brazil's far-right president, Jair Bolsonaro, has been in power. Current estimates suggest areas of rainforest the size of a football pitch are being cleared every single minute.

Reforestation and afforestation are fundamentally limited by the land area available, as trees can only hold a finite amount of carbon. We must also remember that reforestation is not always the best option, and this is why the term 'rewilding' is used in conjunction with reforestation. For example, draining wetlands or peatlands to plant forests is counter-productive, as the carbon storage will be lower and losses of biodiversity will be considerable. So in each region of the world the most appropriate restoration project must be applied. This could be re-wetting wetlands, preserving peatlands, re-growing mangrove forests, or maintaining open grasslands. If an area is suitable for reforestation, decisions have to be made about the most appropriate species for that area in terms of current and future climate and how to increase local biodiversity and other ecological services. One of the criticisms of the Bonn Challenge is that about half the pledges involve large-scale commercial forest plantations. Plantations only lock up carbon while the trees are growing, and much of this is returned when they are harvested. In any case, monoculture is bad for biodiversity.

Geoengineering or technofixes

Geoengineering is the general term used for technologies that could be used to either remove GHGs from the atmosphere or to change the climate of the Earth (see Figure 40). Ideas considered under geoengineering range from the very sensible to the completely mad. Geoengineering is not an alternative to massive reductions in global GHG emissions. Instead, most people see geoengineering solutions as a fall back if we are unable or unwilling to cut GHG emissions quick enough.

Carbon dioxide removal. There are three main approaches to the removal and storage of atmospheric CO_2: biological, physical, and chemical.

(1) *Biological CO_2 removal.* Some researchers try to include reforestation and rewilding in the geoengineering portfolio mainly to make it seem more reasonable. This is not appropriate as these approaches lack the essential engineering requirement. One suggested engineering approach to boost biological uptake of CO_2 is fertilizing the oceans with iron. The late John Martin, an oceanographer, suggested that many of the world's oceans are underproducing because of the lack of vital micronutrients, the most important of which is iron, which allows plants to grow in the surface waters. Marine plants need minute quantities of iron, without which they cannot grow. In most oceans enough iron-rich dust is blown in from the land, but it seems that large areas of the Pacific and Southern Oceans do not receive much dust and thus are barren of iron. Martin suggested that we could fertilize the ocean with iron to stimulate marine productivity. The extra photosynthesis would convert more surface-water CO_2 into organic matter. When the organisms die, the organic matter drops to the bottom of the ocean, taking with it and storing the extra

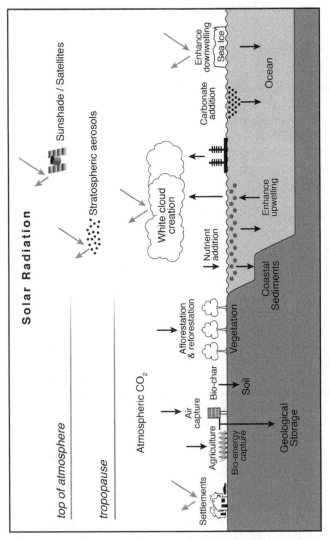

40. The range of geoengineering approaches.

carbon in sediments. The reduced surface-water CO_2 is then replenished by CO_2 from the atmosphere. So, in short, fertilizing the world's oceans could help to remove atmospheric CO_2 and store it in deep-sea sediments. The results of experiments testing the hypothesis at sea have been highly variable, with some showing no effects at all while others have shown that the amount of iron required is huge. The biggest drawback, however, is that if you stop adding extra iron, most of the additional stored CO_2 is released, as very little organic matter is allowed to escape out of the photic zone per year. Ocean fertilization would also have a massive effect on marine ecosystems and biodiversity—because this is deliberate eutrophication on a massive scale.

(2) *Physical CO_2 removal.* It is possible to remove CO_2 directly from the air. However, considering CO_2 makes up just 0.04% of the atmosphere this is much harder and more expensive than it sounds. An early idea was to produce artificial or plastic trees. Klaus Lackner, a theoretical physicist, and Allen Wright, an engineer, supported by Wally Broecker, a climatologist, designed CO_2 binding plastic, which can scrub CO_2 out of the atmosphere. The CO_2 is then released from the plastic and taken away for storage. The first problem is water, as the plastic releases the CO_2 into solution when wet, so the plastic trees would have to be placed in very arid areas or require giant umbrellas. The second problem is the amount of energy required to build and operate these plastic trees and then store the CO_2. The third is one of scale: tens of millions of these giant artificial trees would be required just to deal with US carbon emissions. The advantage over normal trees is that they are not limited to one growth cycle and the CO_2 can theoretically be stored indefinitely. Other technologies to remove CO_2 at source or from the atmosphere are being rapidly developed. For example, Climeworks technology uses giant fans to collect CO_2 directly from the air, producing pure CO_2 to use in industrial processes or even for making artificial fuels. The issue is now one of cost and financing, rather than of engineering.

(3) *Chemical CO₂ removal.* CO_2 is naturally removed from the atmosphere over hundreds and thousands of years through the process of weathering, at a rate of 0.1 GtC per year, but this is a hundred times less than the amount we are emitting. Only weathering of silicate minerals makes a difference to atmospheric CO_2 levels, as weathering of carbonate rocks by carbonic acid returns CO_2 to the atmosphere. By-products of hydrolysis reactions affecting silicate minerals are HCO_3^-, which are metabolized by marine plankton and converted to calcium carbonate. The calcite skeletal remains of the marine biota are ultimately deposited as deep-sea sediments and hence lost from the global biogeochemical carbon cycle for the duration of the lifecycle of the oceanic crust on which they were deposited.

There are a number of geoengineering ideas aimed at enhancing these natural weathering reactions. One suggestion is to add silicate minerals to soils that are used for agriculture. This would remove atmospheric CO_2 and fix it as carbonate minerals and HCO_3^- in solution. The scale at which this would have to be done is very large and there are unknown effects on soils and their fertility. Another suggestion is to enhance the rate of reaction of CO_2 with basalts and olivine rocks in the Earth's crust. Concentrated CO_2 would be injected into the ground and would create carbonates deep underground. An example of this is the Icelandic ON Power geothermal park where CarbFix are taking pure CO_2 provided by Climeworks technology and injecting it into the underground basaltic rock formations. Geothermal renewable energy provides the energy for the direct air capture and injection of the CO_2, and early estimates suggest the system could permanently remove 4,000 tons of CO_2 from the air per year. To put that in context, Iceland would need 1,000 such plants to remove its current annual total carbon emissions. The positive side is that this has now given us a tried, tested, and safe CCS system.

Solar radiation management. Reducing the amount of sunlight hitting or being absorbed by the Earth will reduce the total energy budget and may result in a cooler Earth. As will be evident from the above, some geoengineering solutions are still just ideas and need a lot more work to see if they are even feasible. This is particularly true of the solar radiation management ideas, many of which sound like something out of a bad Hollywood B-movie. These suggestions include changing the albedo (see Chapter 4) of the Earth, to increase the amount of solar energy reflected back into space to balance the heating from global warming (Figure 40). Ideas to increase albedo include erecting massive mirrors in space, injecting aerosols into the atmosphere, making crops more reflective, painting all roofs white, increasing white cloud cover, and covering large areas of the world's deserts with reflective polyethylene-aluminium sheets.

The fundamental problem with all of these approaches is that we have no way of predicting their overall effect on climate. Let us examine the mirrors in space idea put forward by Roger Angel, director of the Centre for Astronomical Adaptive Optics at the University of Arizona. First it would be expensive, requiring 16 trillion gossamer-light spacecraft costing at least $1 trillion and taking 30 years to launch. Second, like all the geoengineering ideas to change the Earth's albedo, it may not work the way we hope it will work. These approaches are aimed at getting the Earth's average temperature down, but they may change the distribution of temperature with latitude, which is what drives climate. Some climate models have shown that these approaches may give a different global climate, with the tropics being 1.5°C colder, the high latitudes 1.5°C warmer, and precipitation varying unpredictably around the world.

Geoengineering governance

One of the major issues with geoengineering is how to govern different groups, companies, and countries playing with the global climate system. There are a great many ethical issues that arise when considering how changing regional and global climate may affect countries differently. There may be overall positive results but minor changes in rainfall patterns, which could mean that whole countries receive too little or too much rain, possibly resulting in disaster. There are three main views on geoengineering: (1) it is a route to buying back some time to allow the UNFCCC negotiations to catch up so that we can achieve net zero carbon by 2050; (2) it represents a dangerous manipulation of the Earth system and may be intrinsically unethical; or (3) it is strictly an insurance policy to support mitigation and adaptation efforts if they fail to be sufficient on their own. Even if research is allowed to go ahead and geoengineering solutions are required, like many emerging areas of modern technology, new flexible governance and regulatory frameworks will be required. Currently there are many international treaties with a bearing on geoengineering, and it seems that no single instrument applies. Hence geoengineering, like climate change, challenges our nation-state view of the world, and new ways of governing will be required in the future.

If we are to solve the problems of climate change, we need to tackle two fundamental issues. The first is how we can reduce to net zero the amount of GHG pollution that we are emitting, while enabling the very poorest countries to develop. The world's population is currently just over 7.8 billion and it is likely to rise and plateau at 10 billion by 2050. That adds up to 8 billion people aspiring to have the same lifestyle as those living in the developed world, with a potentially huge increase in GHG emissions in this century, if they follow the same development pathway to fuel this consumer dream. The second issue is whether as a society we are prepared to invest the relatively small amount, about 1–3% of the world's GDP, to offset a much larger bill in the future. If so, then we have the technology at the moment both to protect our population from climate change and to mitigate the huge predicted emissions of GHGs over the next 80 years. Energy efficiency, renewable energy, CCS, carbon trading, and offsetting all have a role to play. We must also consider 'disruptive technologies', that is, new technologies that we may not yet have even thought of that could change the way we produce or use energy. For example, most of us cannot think of life without a mobile phone or a computer, but this technology has been around for only a few decades, showing how quickly we can become accustomed to change. There are also huge amounts of money to be made from opportunities surrounding changes to our energy use and our personal lifestyles, and, as we will see in Chapter 9, there may be many win-win situations whereby quality of life can be improved at the same time as stabilizing the climate of our planet.

Chapter 9
Changing our future

Introduction

The challenge of climate change must be seen within the current, dominant political and economic landscape. Only by understanding the fundamental societal and economic causes of carbon emissions can we hope to build systems that could rapidly reduce them. At the same time as we deal with climate change we need to ensure that we also tackle other global challenges, such as global poverty/inequality, environmental degradation, and global insecurity. Future policies and international agreements need to provide win-win solutions that deal with the biggest challenges facing humanity in the 21st century.

Planetary stewardship

Scientists struggle with the fact that despite the huge weight of evidence indicating climate change, a small, vocal, but significant, minority of influencers continue to deny that climate change is happening. Scientists have responded by collecting even more evidence. This is called a 'deficit model' response, with scientists presuming that decisions are not being made to mitigate climate change because of a lack of information.

However, social scientists have found that acceptance of climate change has little to do with science and everything to do with politics. Acceptance of climate change represents a challenge to the Anglo-American neoliberal view held by many mainstream economists and politicians. Climate change shows a fundamental failure of the market, and it requires governments to act collectively to regulate industry and business. It is one of the greatest ironies that the very politicians who are denying climate change because of perceived threats to free market values are nonetheless happiest to endorse over $5 trillion of subsidies for the fossil-fuel industry per year. It is a myth that there is any truly free market—many countries happily support subsidies and the blocking of imports.

Neoliberalism encapsulates a set of beliefs that include: the need for markets to be free; for state intervention to be as small as possible; strong private property rights; low taxation; and individualism. Underlying neoliberalism is the assumption that it provides a market-based solution that enables everyone to become wealthier. This so-called trickle-down effect has been the central construct of neoliberalism for the past 40 years but there is no evidence for it happening. Half the world's population lives on less than $5.50 per day. In fact, Oxfam have calculated that the twenty-six richest people in the world currently own the same amount of wealth as those 3.8 billion poorest people collectively. The IMF recently declared that the last generation of economic policies may have been a complete failure.

The global Covid-19 pandemic that started in 2020 has also changed many people's view of neoliberalism. Citizens all around the world have been shown that there can be a different relationship between government, industry, and civil society—a relationship where health and wellbeing are put before economic gain for a country or small minority of individuals. When society faces a real crisis that needs strong coordinated action, it looks to the state and to scientific experts, and for the support of civil society. The private sector can play an important role, such as ensuring food supplies in the face of panic buying or retooling to produce essential medical supplies or to create vaccines. But, equally, many companies look to the state simply for loans and bailouts.

Given the long-term challenges of climate change, biodiversity loss, and possible repeat pandemics, the major lesson from Covid-19 is the failure of free markets to protect us. Instead, it is state intervention, guided by experts, incorporating and valuing society and communities, underpinned by supportive and dynamic business, that is required—to deal with the climate change as well as other challenges of the 21st century. What we need is a new era of planetary stewardship led by governments and underpinned by new economic theories.

Taking action

Potential solutions to reduce global carbon emissions were outlined in Chapter 8, but if we are to reach net zero emissions by 2050 we need to implement all of them. Project Downdraw (at https://drawdown.org/) has identified over eighty high-level solutions that can be implemented at a range of scales to achieve the 1,050 GtC emissions removal required by 2050. Figure 41, based on the work of Avit Bhowmik and colleagues, shows how many of these solutions can be implemented at each level—ranging from the individual to the global. Individual and family action can remove 14% of the required 1,050 GtC, while action at the town and community level can remove 31%, and action at city and state level could remove 33%. This is a counter-blast to the climate change deniers who suggest that it is individuals, not companies or governments, who should take responsibility for dealing with climate change. This shift in blame allows climate change deniers to continue to support the fossil-fuel industry because they argue that it simply meets the demands of the market. Individual and family action is important, as it shows governments and corporations that people are serious about acting to counter climate change, but it is

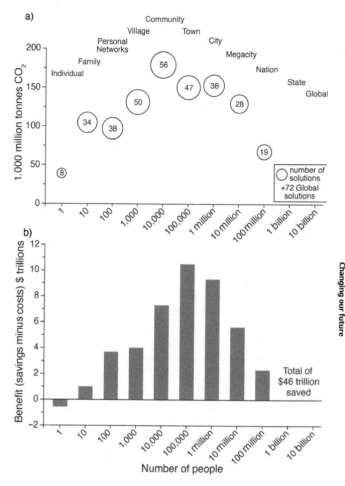

41. Potential climate change solutions from the individual to global scale.

not the solution. Climate change solutions are most effective when carried out from the community to the national level. All of these solutions are win-win and taken together the net benefit (savings minus costs) could be over $46 trillion.

Government, corporations, and civil society

To achieve effective carbon emission reduction requires a partnership between government, both local and national, corporations, and civil society that is supported and encouraged by individual behaviour change.

Governments control the aspirations of civil society through the rule of law and the development of policy. It is clear that governments can use incentives, subsidies, taxation, and regulation to make our societies more sustainable and carbon neutral. Governments are also the major driver of innovation, through investment in university research, funding industrial research and development, and driving demand through incentives. Governments can facilitate the rapid switch from fossil fuels to renewable energy, ensure buildings are carbon neutral, encourage reforestation and rewilding of large areas, promote low-emission farming and more plant-based diets, and support the very poorest people in society, to help build resilience to the likely impacts of climate change.

The world's top hundred companies generate more than $15 trillion in revenue per year. In many ways businesses control our lives as they influence what we eat, what we buy, what we watch, and even who we vote for. Many are already changing, adopting science-based targets so they can achieve net zero carbon emissions by 2050. The challenge for business and industry if they want to remain relevant and trusted in the 21st century is to change their relationship with the environment and society. The classic, linear economic model, 'take, make, dispose', relies on large quantities of cheap, easily accessible materials and energy. We are reaching the physical limits of this model. New inclusive economic theories are emerging showing the fundamental issues with the throwaway corporate culture. The circular economy is essential if companies are to be part of the climate change solution. The circular economy minimizes the amount of resources that are extracted and maximizes the value of products and materials throughout their lifecycle, through reuse and recycling. Applying a circular economy could unlock up to €1.8 trillion in value for Europe's economy. So companies need to plan and make products that have longevity, upgradability, and recyclability built-in. They need to design out waste and pollution.

Though individual actions will only make a small contribution to carbon reduction, they are extremely important as they send a strong message to both government and corporations that citizens want and support major changes. Individual action has had an impact. The School Climate Strikes and the Extinction Rebellion protests have brought together diverse groups of people across the world, all wanting governments to start taking the protecting of our planet seriously. And change is starting to happen, with over 1,400 local governments and over 35 countries having declared that we are in a climate emergency. But we must also remember that not everyone is equally responsible for the current climate crisis: 50% of carbon emissions directly related to lifestyle are emitted by the richest 10% of the world (Figure 42); the poorest 50% of our global society emit just 10% of the pollution. Individual action undertaken by the very wealthiest in society could have a major impact on global carbon emissions.

International institutions

To support, encourage, and when required enforce positive change, we need international organizations fit for the challenges of the 21st century. Many of these institutions, such as the UN, World Bank, IMF, were formed just after the Second World War. Others such as OECD (Organization for Economic Co-operation and Development) and OPEC (Organization of the Petroleum Exporting Countries) were formed in the early 1960s. There is a need for these international institutions to represent everyone in

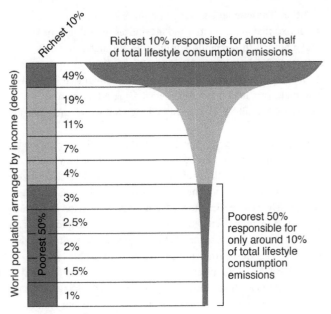

42. Global lifestyle carbon emissions by income group.

the world, and to ensure fair and equitable governance. The
World Bank and IMF could be redesigned so that they focus on
developing the green sustainable economy, supporting the net
zero emission targets, and alleviating poverty, with the Sustainable
Development Goals at the heart of all their decision-making.
The present aim of the WTO is to ensure that trade flows as
smoothly, predictably, and freely as possible, but encouraging
trade and consumption makes reductions in global carbon
emissions harder. It can prevent meaningful local, national, and
international environmental protections and regulations. Perhaps
the WTO could transform into the World Sustainability
Organization (WSO), the first aim of which could be to support
and help restructure economies of countries that rely on fossil-fuel
exports.

One quick and simple change that could be made is to upgrade the UN Environment Agency, because it has a secondary status within the UN system, below that of trade, health, labour; and even of maritime affairs, intellectual property, and tourism. The UN Environment Agency's budget is small, less than a quarter of the UN World Health Organization's (WHO's) budget and a tenth of that of the UN World Food Programme, despite being central to both health and food security. If the UN Environment Agency were to be upgraded to the UN World Environment Organization (WEO), and given a budget at least the size of that of the WHO, it could oversee the Sustainability Development Goals, the Convention on Biological Diversity, and the Convention on Climate Change, to ensure they are mutually reinforcing and not in opposition—making sure there are always win-win-win solutions.

Conclusion

Climate change is one of the few areas of science that makes us examine the whole basis of modern society. It is a subject that has politicians arguing, sets nations against each other, questions the role of companies in society, queries individual choices of lifestyle, and ultimately asks questions about humanity's relationship with the rest of the planet. Only by working together can we deal with one of the greatest crises that has ever faced humanity. There is very little doubt that climate change will accelerate in this century; our best estimates suggest a global mean surface temperature rise of between 2.1°C and 5.5°C by the end of the 21st century. Sea level is projected to rise by between 50 cm and 130 cm by 2100, with significant changes in weather patterns, and more extreme climate events. World leaders have pledged to keep climate change to less than 2°C and if at all possible below 1.5°C. This book has demonstrated that we have the science to understand the causes, consequences, and potential solutions for climate change. We have the technology, the resources, and the money to deal with climate change. What we currently lack are the political will and policies to enable all the positive win-win solutions needed to make a better, safer, healthier, and hopefully happier world. With a growing awareness of the environmental crisis facing the planet, public pressure for change is growing, and new policies and ways of thinking are starting to emerge. The question is whether these changes will be soon enough to get the world to net zero carbon emissions by 2050 (Figure 43).

43. *USA Today* **cartoon of the Copenhagen climate conference.**

Further reading

History of climate change

Corfee-Morlot, J., et al. Climate science in the public sphere, *Philosophical Transactions A of the Royal Society*, 365/1860 (2007): 2741–76.

Leggett, J.K. *The Winning of the Carbon War: Power and Politics on the Front Lines of Climate and Clean Energy* (Crux Publishing, 2018).

Lewis, S.L. and M.A. Maslin *The Human Planet: How Humans Caused the Anthropocene* (Penguin and Yale University Press, 2018).

Mann, M. *The New Climate War: The Fight to Take Back Our Planet* (PublicAffairs, 2021).

Oreskes, N. and M. Conway *Merchants of Doubt: How a Handful of Scientists Obscured the Truth on Issues from Tobacco Smoke to Global Warming* (Bloomsbury, 2012).

Ruddiman, W.F. *Plows, Plagues, and Petroleum: How Humans Took Control of Climate* (Princeton Science Library, 2016)

Weart, S.R. *The Discovery of Global Warming, New Histories of Science, Technology, and Medicine* (Harvard University Press, 2008).

Science

Archer, D. *Global Warming: Understanding the Forecast*, 2nd edn (John Wiley & Sons, 2011).

Dessler, A.E. *The Science and Politics of Global Climate Change: A Guide to the Debate*, 3rd edn (Cambridge University Press, 2019).

Emanuel, K. *What We Know about Climate Change*, updated edn (The MIT Press, 2018).

Houghton, J.T. *Global Warming: The Complete Briefing*, 5th edn (Cambridge University Press, 2015).

IPCC, Climate Change 2021—The Physical Science Basis Contribution of Working Group I to the Sixth Assessment Report of the Intergovernmental Panel on Climate Change (2021).

Lenton, T. *Earth System Science: A Very Short Introduction* (OUP, 2016).

Maslin, M. The five corrupt pillars of climate change denial (The Conversation, 2019). https://theconversation.com/the-five-corrupt-pillars-of-climate-change-denial-122893

Maslin, M. Five climate change science misconceptions—debunked (The Conversation, 2019). https://theconversation.com/five-climate-change-science-misconceptions-debunked-122570

Maslin, M.A. and S. Randalls (eds) *Routledge Major Work Collection: Future Climate Change: Critical Concepts in the Environment* (4 volumes containing reproductions of eighty-five of the most important papers published on climate change) (Routledge, 2012).

National Climate Assessment. Volume I: Climate Science Special Report (2018). https://science2017.globalchange.gov/

Romm, J. *Climate Change: What Everyone Needs to Know* (OUP, 2018).

Impacts

Costello, A., et al. Managing the health effects of climate change, *The Lancet*, 373 (2009): 1693–733.

Garcia R.A., et al. Multiple dimensions of climate change and their implications for biodiversity, *Science*, 344 (2014): 486–96.

IPCC, Climate Change 2021—Impacts, Adaptation, and Vulnerability, Contribution of Working Group II to the Sixth Assessment Report of the Intergovernmental Panel on Climate Change

National Climate Assessment. Volume II: Impacts, Risks, and Adaptation in the United States (2018) https://nca2018.globalchange.gov/

Stern, N. *The Economics of Climate Change: The Stern Review* (Cambridge University Press, 2007).

Watts, N., et al., The 2020 Report of The Lancet Countdown on Health and Climate Change (The Lancet, 2020).

Politics and governance

Figueres, C. and T. Rivett-Carnac. *The Future We Choose: Surviving the Climate Crisis* (Manilla Press, 2020).

Giddens, A. *The Politics of Climate Change*, 2nd edn (Polity Press, 2011).

Grubb, M. *Planetary Economics: Energy, Climate Change and the Three Domains of Sustainable Development* (Routledge, 2014).

Gupta, J. *The History of Global Climate Governance* (Cambridge University Press, 2014).

IPCC, Climate Change 2022—Mitigation of Climate Change, Contribution of Working Group III to the Sixth Assessment Report of the Intergovernmental Panel on Climate Change.

Klein, N. *On Fire: The Burning Case for a Green New Deal* (Allen Lane, 2019).

Labatt S. and R.R. White *Carbon Finance* (Wiley, 2007).

Metcalf, G.E. *Paying for Pollution: Why a Carbon Tax is Good for America* (OUP, 2019).

Meyer, A. *Contraction and Convergence: The Global Solution to Climate Change* (Green Books, 2015).

Oxfam, Policy Paper—Confronting Carbon Inequality: Putting climate justice at the heart of the COVID-19 recovery (Oxfam, 2020). https://oxfamilibrary.openrepository.com/bitstream/handle/10546/621052/mb-confronting-carbon-inequality-210920-en.pdf

Thunberg, G. *No One Is Too Small to Make a Difference*, paperback (Penguin, 2019).

Solutions

Buck, H.J. *After Geoengineering: Climate Tragedy, Repair, and Restoration* (Verso, 2019).

Centre for Alternative Technology (CAT), *Zero Carbon Britain: Rising to the Climate Emergency* (CAT Publications, 2019). https://www.cat.org.uk/new-report-zero-carbon-britain-rising-to-the-climate-emergency/

Cole, L. *Who Cares Wins: Reason for Optimism in Our Changing World* (Penguin, 2020).

Georgeson, L., M. Poessinouw, and M. Maslin *Assessing the Definition and Measurement of the Global Green Economy* (Geo: Geography and Environment, 2017). doi: 10.1002/geo2.36

Goodall, C. *What We Need to Do Now: For a Zero Carbon Future* (Profile Books, 2020).

Hawken, P. *Drawdown: The Most Comprehensive Plan Ever Proposed to Reverse Global Warming* (Penguin, 2018).

Helm, D. *Net Zero: How We Stop Causing Climate Change* (William Collins, 2020).

IPCC, Climate Change 2021—Impacts, Adaptation, and Vulnerability, Contribution of Working Group II to the Sixth Assessment Report of the Intergovernmental Panel on Climate Change.

Jackson, T. *Prosperity without Growth: Economics for a Finite Planet* (Routledge, 2016).

Maslin, M. Stabilising the global population is not a solution to the climate emergency (The Conversation, 2019). https://theconversation.com/stabilising-the-global-population-is-not-a-solution-to-the-climate-emergency-but-we-should-do-it-anyway-126446

Morton, O. *The Planet Remade: How Geoengineering Could Change the World* (Granta, 2016).

Roaf, S., et al. *Adapting Building and Cities for Climate Change* (Routledge, 2009).

Royal Society, Geoengineering the climate: Science, governance and uncertainty: The Royal Society Science Policy Centre Report, *The Royal Society*, 10/09 (2009): 81.

General reading

Berners-Lee, M. *There Is No Planet B: A Handbook for the Make or Break Years* (Cambridge University Press, 2019).

Flannery, T. *Atmosphere of Hope: Solutions to the Climate Crisis* (Penguin, 2015).

Hayhoe, K. *The Answer to Climate Change: And Why We Can Have Hope* (Atria/One Signal Publishers, 2021).

Lynas, M. *Our Final Warning: Six Degrees of Climate Emergency* (Fourth Estate, 2020).

Mazzucato, M. *The Value of Everything: Making and Taking in the Global Economy* (Penguin 2019).

Raworth, K. *Doughnut Economics: Seven Ways to Think Like a 21st-Century Economist* (Random House, 2017).

Royal Society, People and the planet, The Royal Society Science Policy Centre Report, *The Royal Society*, 01/12 (2012): 81.

Sachs, J. *The Ages of Globalization* (Columbia University Press, 2020).

Wallace-Wells, D. *The Uninhabitable Earth: A Story of the Future* (Penguin, 2019).

Index

Climate Change

DESERTS
A Very Short Introduction
Nick Middleton

Deserts make up a third of the planet's land surface, but if you picture a desert, what comes to mind? A wasteland? A drought? A place devoid of all life forms? Deserts are remarkable places. Typified by drought and extremes of temperature, they can be harsh and hostile; but many deserts are also spectacularly beautiful, and on occasion teem with life. Nick Middleton explores how each desert is unique: through fantastic life forms, extraordinary scenery, and ingenious human adaptations. He demonstrates a desert's immense natural beauty, its rich biodiversity, and uncovers a long history of successful human occupation. This *Very Short Introduction* tells you everything you ever wanted to know about these extraordinary places and captures their importance in the working of our planet.

www.oup.com/vsi

GEOGRAPHY
A Very Short Introduction
John A. Matthews & David T. Herbert

Modern Geography has come a long way from its historical roots in exploring foreign lands, and simply mapping and naming the regions of the world. Spanning both physical and human Geography, the discipline today is unique as a subject which can bridge the divide between the sciences and the humanities, and between the environment and our society. Using wide-ranging examples from global warming and oil, to urbanization and ethnicity, this *Very Short Introduction* paints a broad picture of the current state of Geography, its subject matter, concepts and methods, and its strengths and controversies. The book's conclusion is no less than a manifesto for Geography' future.

> 'Matthews and Herbert's book is written- as befits the VSI series- in an accessible prose style and is peppered with attractive and understandable images, graphs and tables.'
>
> **Geographical.**

GEOPOLITICS
A Very Short Introduction
Klaus Dodds

In certain places such as Iraq or Lebanon, moving a few
feet either side of a territorial boundary can be a matter of life
or death, dramatically highlighting the connections between
place and politics. For a country's location and size as well as
its sovereignty and resources all affect how the people that live
there understand and interact with the wider world. Using
wide-ranging examples, from historical maps to James Bond
films and the rhetoric of political leaders like Churchill and
George W. Bush, this Very Short Introduction shows why,
for a full understanding of contemporary global politics, it is
not just smart - it is essential - to be geopolitical.

'Engrossing study of a complex topic.'

Mick Herron, Geographical.

LANDSCAPES AND GEOMORPHOLOGY
A Very Short Introduction
Andrew Goudie & Heather Viles

Landscapes are all around us, but most of us know very little about how they have developed, what goes on in them, and how they react to changing climates, tectonics and human activities. Examining what landscape is, and how we use a range of ideas and techniques to study it, Andrew Goudie and Heather Viles demonstrate how geomorphologists have built on classic methods pioneered by some great 19th century scientists to examine our Earth. Using examples from around the world, including New Zealand, the Tibetan Plateau, and the deserts of the Middle East, they examine some of the key controls on landscape today such as tectonics and climate, as well as humans and the living world.

www.oup.com/vsi